THE TRUE STORY OF CATCH-22

"Never has a great work of art like *Red Badge of Courage* or *For Whom the Bell Tolls* or *War and Peace* been so closely tied to real-life individuals and real-life events.

Literary scholars will be pouring over your book ... for centuries to come looking for clues to the genius of *Catch-22*".

-Daniel Setzer, 57th Bomb Wing historian/archivist

The True Story of
Catch-22

The Reality that Inspired
one of the Great Classics
in American Literature

Illustrated Edition

The True Story of Catch-22 is the authoritative companion to
Joseph Heller's classic *Catch-22*.

This edition is enriched by the inclusion of a bold center section of full-color illustrations of *Catch-22's* factitious characters contrasting with their true-life wartime counterparts.

Also introduced are films, untouched since WWII ended, that were discovered recently deep in the National Archives. From this footage, copious never before published stills are here extracted of the 340th Bomb Group.

Astonishingly they contain substantial, rare, and touching appearances by
Joseph Heller - warrior before writer.

Patricia Chapman Meder

CONTENTS

DEDICATION

My father, Bill, my model for setting goals and accomplishing them.
My mother, Charlotte, gracious, elegant, and always in my corner.
My sister, Sue, the other half of my heart.
My son, Jason, brilliant and compassionate; a game changer.
My husband, Jim, my rock solid and loving support throughout everything.

FOREWORD BY SCOTT CARPENTER

Among the vast wealth of literature that emerged from World War II, none has stood the test-of-time better than Joseph Heller's first novel, *Catch-22*. A black-comic masterpiece, written from firsthand experience, it has resonated with half a century of readers since, providing timeless insights into the human psyche.

The men of Joseph Heller's 340th Bomb Group had very serious and dangerous jobs to do, and it is this that makes these same men and events such fertile grounds for satire. They flew 898 combat missions in B-25 Mitchell bombers until the surrender of Germany in 1945. Every time they boarded their aircraft, there was a very real risk of not returning. In my experience as a naval aviator and Mercury astronaut I am familiar with the risks they took flying into the unknown.

Heller himself was a kid from Brooklyn, when he suddenly found himself the bombardier on a B-25 Mitchell based in Italy. Surrounding him were farm boys, shoemakers, factory workers and prospective teachers. But now they were flyers in combat with their German counterparts—most likely also naive young men of different backgrounds—trying to shoot each other down.

Joseph Heller was able to identify the immense ironies in his situation, and *Catch-22* captured the absurdities from the start. As I read *Catch-22*, I respected the humor. I also respected the difficulty of the training for such dangerous situations that was required for the accomplishment of the mission. The humor was based largely on the desire to step away from the complexity and stress of those missions and mitigate the feelings that each mission might be one's last.

Along with me in the Mercury Seven was Deke Slayton. Deke had flown 56 combat missions in Europe with the 340th Bomb Group. Eventually he and his war comrades were able to laugh at and respect the humor of *Catch- 22*, but because of the seriousness of their roles in World War II, they did not always appreciate it.

In interviews after his novel was published, Heller sometimes denied that the characters in his work were based on real persons. And here is where the current work, *The True Story of Catch-22*, provides an invaluable service to all readers of World War II literature. Patricia Chapman Meder, daughter of the commander of Heller's 340th Bomb Group, has intensely researched her father's wartime career, corresponded with other intimately involved veterans, and Heller himself, to solidly establish that *Catch-22* was indeed based on reality.

The results of her research in these vividly written pages, has been revelatory. *Catch-22* was by no means a flight of Heller's imagination, but in very large part the true experience of our citizen-soldiers who were suddenly assigned combat tasks the world had not seen before. The real-life story of the men in Heller's Bomb Group is more fascinating than the novel. After all these years we now find that the novel was based on real people and not just caricatures or figments of imagination.

In real life, the courage, and humor, of our young flyers in World War II stands as even more impressive than their fictional depictions. Heller's sharp satiric eye has been a subject of fascination for generations. Let's hope that the story of the real young men from whom he gained his inspiration are also kept in mind, for it was their own camaraderie, idiosyncrasies, and courage at that time of deep patriotic need that made the original novel possible. In these pages it is a great pleasure to finally see the real story behind the fictionalized account, and to be even more impressed.

Fact is indeed better than fiction and this book is a great read about fact.

Scott Carpenter
Mercury Astronaut
Aurora Seven
24 May, 1962

INTRODUCTION

I slid into the passenger's seat of his new red 1962 VW and we set out for the 6th *arrondissement* in Paris. He eased into the traffic, quickly adapting to the French method of survival driving as a slight *motocyclette* carrying a bouffant-haired, mini-skirted *femme*, perched sidesaddle and plastered bosom-to-back to whoever was in front, darted around us. He pointed to the small, recently published, book between us.

"Take a look at what I'm reading; just read the first page."

As the opening sentence warned, "It was love at first sight."

A few days later he loaned it to me with a bit of a grin saying, "I think this might be your Dad's World War II outfit." And there began my lengthy on-again, off-again journey with the phenomenal *Catch-22.*

I came from a military family—actually, generations of military family—and my older sister, Sue, and I grew up with its expectations, inconveniences, privileges, and responsibilities. From my earliest memories the effects of WWII have been a part of my life. I can remember toying with my food and Dad, zeroing in on me with spoon raised high, saying "Bombs away!" or "Open the bomb bay!" or "Down the hatch!" Having been alive a mere four years, I was clueless to the phrases, of course, but I knew the expectations: Open wide and the food will make a direct hit.

Over the years I have absorbed, almost by osmosis, stories and accounts by my father, final commander during World War II of the 340th Bomb Group, a senior pilot, a rated navigator, a rated bombardier, and an observer. Upon undertaking this book, I sifted through his copious, museum-worthy war materials; much of it formerly classified Secret. I interviewed the men involved in the now-famous 340th Bomb Group depicted in *Catch-22,* published by Simon & Schuster, 1961, especially those from Heller's 488th squadron. I went to reunions and I read and listened to first-person story after story. I corresponded with Joseph Heller. I learned a great deal as the years passed.

The 340th Bomb Group was comprised of four squadrons: the 486th, 487th, 488th, and 489th. This group, in turn, was one of three that came under the command of the 57th Bomb Wing, a wing that was, in a few short years, to be the recipient of unexpected notoriety due solely to one unnoticed bombardier.

This bombardier, Lieutenant Joseph Heller, had joined the 340th, which was based in Corsica, in May 1944, just two months after my father, Col. Willis F. (Bill) Chapman, became its last commander. Four days later, nineteen-year-old Heller, a combat replacement, was assigned to his first mission as a wing bombardier. From this moment in his young life until the war ended sixteen months later, Heller was immersed in the demands of this deadly conflict.

In 1944, as the Allied Forces in Italy pressed toward the war's European resolution, the US 488th Bomb Squadron kept doggedly to its mission. The lives of these men of the 488th were following the dictates of the war; there was information to absorb, skills to learn and hone, mind-sets to adjust, fears to conquer, and, of course, missions to fly.

What they were unaware of was that, as they focused on their daily lives, one of their members was focusing upon them. He was absorbing a different type of information; information that was to serve him well years after the war had claimed its last son and daughter. Keeping all of his senses alert to his surroundings, 488th bombardier Joe Heller was absorbing impressions that would later astonish the literary world when he penned a phenomenal novel that exploded across the planet in the 1960s. Welcome, world, to *Catch-22*!

What had begun harmlessly enough with a young man's eagerness to experience heroic adventures far too quickly matured for Heller into the never-to-be-forgotten, life-altering experiences only war can create. He was a good soldier; he did as he was told. He was one of the fortunate who survived the ordeal. And then he shipped home with a wealth of memories —of missions, of crewmates, of experiences.

Nine years later he began to write. This richness of memories, of knowledge that he had absorbed, fueled his pages. In varying degrees his war-mates morphed into wildly factitious characters trapped in the chaos of *Catch-22* while deadly missions became . . . well, still, deadly missions.

In most of Heller's interviews he chose to deny similarities between his clever fictional characters and his war-mates; however, in other interviews, he would occasionally admit to basing certain characters on real-life counterparts. But to the men of the 340th, Heller's offspring were instantly recognizable. Aside from their positions and ranks, men were easily identifiable by their descriptions, mannerisms, and traits. In a wild and imaginative frenzy, characters such as the hard-drinking, vengeful, and disillusioned Chief White Half Oat; young, sliced-in-half Kid Sampson; shrieking, frenzied Hungry Joe, and a slew of others blasted skyward from the solid roots of real life bombardier Vincent Myers, pilot Bill Sampson, pilot Joseph Chrenko—and the list goes on. Not only could the origins of various characters be proven, but the origins of actual situations could as well.

As certain as a perfect chicken's egg is able to become the burnt-topped crème brûlée, slippery-slick and tissue-thin-crisp, so, under chef Joseph Heller's creative spoon, were his crewmates able to be brilliantly remixed and reshaped. Four of these unsuspecting men were to become four of *Catch-22*'s major characters. Witness how true-life George, Bill, Bob, and creator Joe funneled into the loopy, rogue, and ripe-to-bursting factitious personalities of the humble, capable Capt. Wren, the naked-in-the-ranks Lt. Yossarian, the "Black-Eye" or "Feather-in-my-Cap" Col. Cathcart, and the arbitrary, unpredictable Gen. Dreedle.

With the finesse of a Formula One driver, Heller slid, cornered, and shifted characters from fact into fiction throughout the growth of his novel. Major Jerre Cover eased into Major __de Coverley, while Douglas Orr actually retained his own name as he morphed into a new persona. Major Major accelerated into Major Major Major Major, and even the dog, "to protect its identity," mutated into a cat. The precision of Heller's touch was masterful. But in the end, even though Heller's fully developed characters stand solely, solidly, and uniquely on their own merits, still it must be acknowledged that any resemblance to persons living or dead is, in fact, actual.

The True Story of Catch-22 is divided into thirds. Part I highlights four solid Air Force officers whom Joseph Heller blindsided when he creatively massaged them into four of *Catch-22*'s heavy hitters. But to paraphrase slightly, "Truth is as fascinating as fiction." Part II rewards our brain's right side with full-color illustrations of *Catch-22*'s factitious spawn contrasting strongly, on facing pages, with our actual WWII heroes brought to life in previously unpublished photos and accompanied by first-person narratives. Lastly, Part III goes to the heart of this book as twelve men of the 340th relate twelve true tales. In miniature, here lies the true story.

The telling of *The True Story of Catch-22* is a major responsibility for me and I here state that, as General William T. Sherman of Civil War renown wrote in his recollections, "Of omissions there are plenty, but of willful perversion of facts, none."

My regret in this venture is the passing of time. Now nearly all the characters upon whom Heller drew for his masterpiece are gone—Cathcart/ Chapman, Dreedle/Knapp, Piltchard/Dyer, Daneeka/Marino, Havermyer/ Myer, Major Major Major Major/ Major Major, Tappman/Cooper, and the rest. All have passed now.

All but one: Wells/Wren.

It is Captain George L. Wells, an extraordinary pilot, who will accompany me through these pages. From his first mission, October 26, 1943, to his last on March 19, 1945, George kept a small black mission book, now well thumbed. Every one of his historic 102 missions was accurately recorded and, with the regularity of almost daily flights, each of these missions will keep step with the pages of this book.

As familiar as I am with things military, it was, nevertheless, while quietly, privately reading the brief chronological entries in George's mission book, that I received the most profound insight into a bomber pilot's life as it daily unfolded. I could feel this war.

It has now been more than half a century since *Catch-22*'s publication. Currently, the surviving men of the 57th Bomb Wing, which was comprised of the 310th, 319th, 321st, and Heller's 340th Bomb Groups, while celebrating at their annual WWII reunions, will still chuckle or grimace, discuss or deride the novel—the book that, no matter whether it had become an American classic, no matter how clever or unique, still had chosen to poke huge fun at a circumstance that took a monumental toll on the lives of these men.

While they acknowledge and can even respect the profound impact of *Catch-22*, still, it is to a measure at the expense of those who steadily had put their lives on the line, as had Joe Heller, mission after mission, for a cause of such magnitude. They find it difficult to fully embrace the levity even decades after the fact. It is for them that I decided to write *The True Story of Catch-22*.

— Patricia Chapman Meder

"IF YOU WERE TO ASK ME …"

From Joseph W. Ruebel—the seed of Catch-22's Major Danby, the timid, soft-spoken operations officer and former college professor who sees himself as a poor fit for the armed services but does his best for his country:

"If Heller had asked me if I wanted to be in the book I'd say OK. All right, I guess I'm honored to be in it. It was a heck of a good idea. A marvelous idea. Initially the book bothered and irritated me since the facts were so distorted. Now I am more mellow. The more I thought about it the funnier it got.

"I loved that B-25. In the year I was there, there was only one B-25 accident due not to combat. That occurred when the pilot mistook a golf course for the aerodrome."

From Robert D. Knapp—the seed of Catch-22's General Dreedle, the blunt, ill-tempered old general who knew he was the hammer and everyone else was a nail:

"I have no use for anybody who would take the best Bomb Wing in the world and make caricatures of its men. Anybody can write derogatorily if he wants to. The question is, why would he want to? Those he wrote about were a fine bunch. All earned awards and citations. We had a great outfit. I am proud of all of them."

From Dorothy Knapp Spain, Robert Knapp's daughter:
"My dad hates the book."

From George L. Wells—the seed of Catch-22's Captain Wren, the gentle, quiet squadron operations officer who, in tandem with Capt. Piltchard, assigned missions impartially to himself and his crewmates for all to enjoy:

"Of course I don't mind being in *Catch-22*. It has been a plus for me.

In 1993, while working on my house, I was hurt in a bad fall. During the night I got up for a glass of water and fainted. The following morning I was checked into the hospital to be examined for a stroke or heart attack. I went through a battery of tests. One of these involved a visit by the psychiatrist to check for possible brain damage. In his series of questions he asked, 'Have you read *Catch-22*?' When I replied, 'I'm in the book,' his immediate response was, 'Well, I HAVE got a nut here!' "

From Willis F. Chapman—the seed of Catch-22's Colonel Cathcart, the obsessive, ambitious, mission-raising group commander endowed with indecisiveness and a serious lack of sound judgment:

"I remember being somewhat irritated when the book was first published and I don't believe I ever got through it completely. The military, at that time, was so sensitive about presenting the proper image to the American public. They were very, very serious about this. I kind of lost my sense of humor in Heller's version.

I'm not sure I can regard it a high compliment that I am considered the genesis of the character which Heller developed into Col. Cathcart; however, my feelings are very mixed now that it has given my daughter the incentive to write this book, which I think Heller's war-time comrades might enjoy."

PART I

GEORGE, JOSEPH HELLER,
AND
"THE BEST DAMNED BOMB GROUP THERE IS"

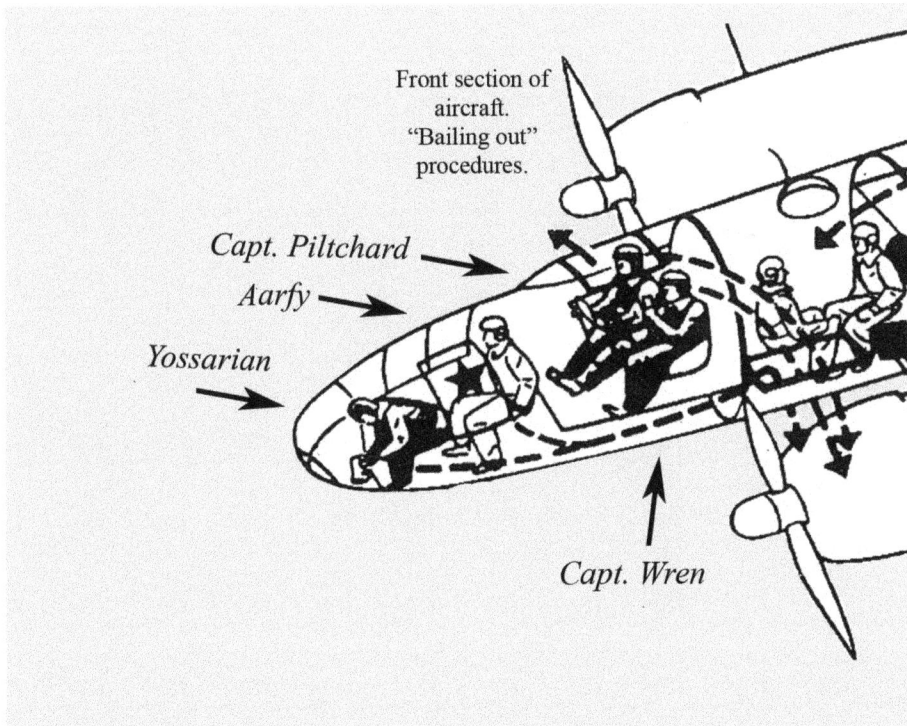

Front section of aircraft. "Bailing out" procedures.

Capt. Piltchard

Aarfy

Yossarian

Capt. Wren

CHAPTER **1**

GEORGE: UNWITTING ANCESTOR OF **CAPTAIN WREN**

At this book's writing George bears some wounds from being 91 years old, but none of them are from having served in World War II. This is a phenomenal fact given that he, along with Fred Dyer, another pilot from his 488th squadron, holds the unofficial U.S. record, at 102 each, for the number of bombing missions flown in the history of that grim war. Both survived the war but, between them, they had been shot at and hit countless times, blasted by missiles, shrapnel, and gunfire. They survived bailouts and crash landings and lived to fly again and again, more than doubling the required number of missions, first set at 25, then raised to 35, and finally plateauing at 50. Their piloting skills were exceptional, their devotion to duty far exceeded expectations, and their bravery was beyond question.

There is another interesting fact about George. Remarkably, he is the last surviving Air Force member from whom many of the characters in Joseph Heller's classic novel, *Catch-22*, were loosely drawn. He is George L. Wells and he is, to a degree, also Captain Wren.

George L. Wells, holding his well-worn black book into which he recorded each of his 102 missions.

Sandy-haired and freckled, George, a child of the Depression, was raised in small town USA. Cedar Brook, New Jersey sported a population of less than 300. His family, living about a mile outside of town, fared well for country folk prior to the Depression. He did all of the things boys do —attended school, played ball (especially baseball and soccer), became his 1937 senior class president, graduated, and got a job.

Since his parents could not afford to send him to college, he started working as an apprentice in a print shop for $8.00 a week. After he'd worked for two years, union organizers came in and told him his salary would increase to $16.00 a week if the print shop employees voted to join the union. Of course the employees voted for it but the print shop, in order to be able to accommodate the increased wages, laid off the last two persons hired, one of whom was George.

In 1939, he was alerted by his older brother, Oliver, that his country was headed for war. On his advice, George joined the New Jersey National Guard and, with war drums beating, his life abruptly reeled into totally unfamiliar territory.

The world's military pot had been simmering but now it was roiling, bubbling, and boiling. Germany, Italy, and Japan were throwing massive logs on the fire with their passion to increase their power and force their expansions at the expense of neighboring countries.

As Adolf Hitler moved into power he promised his country an end to the humiliating conditions caused by World War I. He began re-creating and preparing the German army for a war of conquest. Here was rising a man, unimposing at 5'9" and 150 pounds, with a flattened haircut, a scrap of a moustache, and dead-serious dark eyes. A father-thwarted artist's passion had simmered inside of him but now he was mutating into a fanatical and tyrannical alpha male who in the ensuing few years would declare, "Brutality creates respect." "Go ahead," he commanded, "kill without mercy. After all, who remembers today the Armenian Genocide." He was to decree that the Nazi Party "should not become a constable of public opinion, but it must dominate it. It must not become a servant of the masses, but their master!"

Looking eastward, Hitler's empire was to be in Eastern Europe and the Soviet Union. His goal was the unification of all German-speaking people into one great nation showcased by his created populace of tall, blond, blue-eyed, athletically fit Aryans. He wished to eliminate the weak and the medically handicapped and to racially cleanse the continent. If Hitler were not Hitler, this short, dark-haired man just might have had himself gassed.

His astonishing cunning and deceit in fabricating and justifying his actions allowed him to successfully begin taking control. *Der Führer* was closing his fist and continuing to force expansion, first to Austria, then Czechoslovakia; eventually Denmark, Norway, the Netherlands, Belgium, and France would join the long list. With unsuspecting Poland about to be the third recipient of the Nazi's crushing boot, Great Britain and France rose to meet their destiny and declared war on Germany on Sept. 3, 1939. World War II had begun.

In the United States the people favored neutrality. Recovering from the First World War, the democracies of the world craved peace and thus were totally unprepared militarily. As Americans followed the early march of German occupation as it steadfastly bulled forward, they began getting nervous. Gingerly they started the buildup of their own limited military resources.

President Franklin D. Roosevelt had first called on the US to supply the Allies with badly needed materials. But now the US focus began to turn homeward. Factories converted from sewing machines to machineguns. As America's men flocked to the armed forces, America's women left the comfort of their homes to step into the vacated places in the plants. Officials soon discovered that women could solidly perform the duties of 8 out of every 10 jobs normally done by men.

Meanwhile the German armies took on the Big Bear of Russia, who slashed back savagely through its icy weather. With no winter clothing and vehicles ill-designed for such a severe and punishing climate, Hitler's soldiers were crippled and freezing in Moscow's bottom-dwelling temperatures of twenty to fifty below zero.

As the Germans failed in their attempt to capture that capital, Japan suddenly pushed the US into the conflict with its staggering assault on the unaware and unprepared American naval base at Pearl Harbor, Hawaii. The base, attending its daily duties on that island of pineapples and leis, was stunningly dive-bombed by 353 Japanese aircraft. Four US Navy battleships were sunk and four others were damaged, as well as three cruisers, three destroyers, an anti-aircraft training ship, and one minelayer. 2,402 men were killed and another 1,282 wounded.

The shock to the American people was profound. The following day, Dec. 7, 1941, the day President FDR would later refer to as "a date which will live in infamy," the US declared war on Japan.

FDR called for the immediate and massive expansion of the armed forces. The 20 years of neglect and indifference could not be overcome in a few days. US industry staggered and strained to support its nation's allies with equipment as well as, now, providing for its own military expansion.

And American young men and women stepped up to the plate.

When his National Guard Field Artillery Regiment was called to active duty in 1940, George left boyhood behind. Military schooling, advanced pilot training, which included a crash with his instructor that landed him in the hospital for three weeks, and assignments followed. On October 6, 1943 George boarded the *Empress of Scotland* (the old *Empress of Japan*) and the following day at 11:40 a.m. this homegrown young man sailed into the unknown. Eight days later they docked at Casablanca, French Morocco and another eight days found him in Algeria, Sicily, Italy, Sicily again, then Italy again. George was transforming from country boy to warrior jet setter.

George's maiden mission was plotted; the assigned drab-green B-25 standing by on the airstrip was primed and heavy with its fearsome bombs loaded, and George's whole being keenly revved in high gear. However . . .

MISSION #1: OCT. 26, 1943
> *Bomb the town of Terracina, West Coast of Italy.*
> *Oct. 27 Bad weather—mission cancelled.*
> *Oct. 28 Bad weather—mission cancelled.*
> *Oct. 29 Bad weather—mission cancelled.*
> *Oct. 30 Bad weather—mission cancelled.*
> *Oct. 31   24 B-25s & Kitty Hawk (P-40s) Fighters. Was #2 in 3rd*
> *box of airplanes. After flying to initial point, was called off*
> *due to weather. Target had been harbor at Ancona,*
> *East Coast of Italy.*

Slow, very slow, start . . . . By his fourth mission, however, he had a taste of how his future was going to unfold:

MISSION #4: NOV. 12, 1943

> *To a Tatoi Airport, Athens, Greece—48 airplanes from 340th & 48 from 321st & 36 from 310th, 82nd fighter Wing for support—P-38s, P-39s, P-40s & Spitfires. Never saw so many planes in the air at one time. Weather stopped us from going to original target. Started a run on alternate target on airport at Berat, Albania. The Jerries (Germans) put up flak so thick you could have played baseball on it. All of the old flyers said it was the most ack-ack they had ever seen. Enemy fighters were diving down on gun crews. The other squadrons dropped their bombs but we couldn't get near the target. As yet I don't know whether we got credit for the mission or not. Had 12 frog (smaller) bombs per plane. Finally got credit for mission.*

In June of 1944, somewhat older and infinitely wiser, now battle-accomplished George was moved up to the US Army Air Corps' 340th Bomb Group based on the Mediterranean island of Corsica, a picturesque region of France. Here, on its airfield that had been scraped out of a briarroot patch parallel to the pristine, water-caressed shore, was where he was to remain until war's end.

This bomb group, the notorious "Unlucky 340th," was where eventually crossed the paths of Capt. George L. Wells (Captain Wren), Col. Willis F. Chapman (Colonel Cathcart), General Robert D. Knapp (General Dreedle), Lt. Joseph Heller (Yossarian) and other men soon to be immortalized in literature. Here, in this bomb group, was where events began to unfold that would eventually thread their way through the pages of *Catch-22*.

George had collected five months of hard-earned experience before Col. Chapman took command of the 340th, which was then based at the foot of Mt. Vesuvius in Pompeii, Italy. The original sculptors of this airfield had been British engineers who had whittled it out of a lush grape vineyard.

MISSION #50: MARCH 20, 1944

Led Sqd. flight on target at Perugia, but bad weather made us turn and bomb alternate at Terni. Vesuvius is really acting up and lava is running down the sides of the mountain.

With disbelief and great curiosity, Chapman arrived just in time to experience one very rare and very mammoth volcanic eruption as it hurled and spewed its molten guts directly over their airfield. The thundering, sky-splitting explosion proceeded to bury everyone and everything under a foot and a half of black, basketball-sized rock, clinkers, and ash. It was as if Vesuvius the Magnificent had just been waiting to display its awesomeness. The force of its punch was of such magnitude that bombs and flak seemed the merest of play toys by comparison.

Two months later Joseph Heller arrived, perhaps grateful to miss this scarring event that he would have relished.

Both George and Joe were assigned to the 488th squadron, one of four under Chapman's command. The 340th, in turn, was one of four groups under the leadership of General Knapp, a commander whose military career already had included World War I.

By this time, George Wells knew war. He knew the routine of rising in the early dark of pre-dawn to fill a basin with cold water, wash, shave, and down a cup of coffee with bread, all before hurrying off to the briefing room prior to flying another mission. He knew the sharp metallic crack of the truck tailgate dropping, rattling and clanking open to deposit the officers it had transported from that briefing's coffee to the airstrip in front of a waiting B-25, which, like its crew, had been warmed up. And he knew, as he knew his own heartbeat, the sound, the sight, and the smell of that plane: that B-25 that was both dreaded as the deliverer to death's door, trusted as their dearest friend under unforgiving fire, and worshipped, often as wounded as they were, as their sole means of a safe return home.

Now, so many years later, George also knows every page of *Catch-22*. Author Joseph Heller's ultimately classic novel (confuting Mark Twain's definition of "a book which people praise but don't read") was born of those circumstances and took on a life of its own. George quickly recognized this book was his group's life, his life. And, as the pages turned, he vacillated between clenched teeth and an insuppressible grin. This book belongs to the world but, more importantly, it belongs to him and his wartime buddies.

At the opening of the irreverent and farcical *Catch-22,* Heller sets the stage.

> "The island of Pianosa lies in the Mediterranean Sea eight miles south of Elba. It is very small and obviously could not accommodate all of the actions described." *(Catch-22, p. 6)*

Almost spot-on. The actual setting for this tale was a mere thirty-four miles southwest of Pianosa on the exquisite island of Corsica. This island with its wild and jagged mountains, its deep, shadow-covered valleys, and its clear, cool mountain streams was formed through volcanic explosions and, while it was geographically closer to Italy, it was a region belonging to France. The island enjoyed a heritage of the temperate climate of the Mediterranean. The days of its coastal summers were hot and dry while slowly arriving winters eased in pleasantly mild and rainy relief, frequently evolving into heavy rainstorms. In spite of its very turbulent past, Corsica seemed created to be a place to enjoy life as pristine, clear waters gently lapped its eastern coastline. This was not a place to be touched by war. But here was where George, Joe, their 340th Bomb Group, and 16 other United States military airfields were located.

The umbrella 57th Bomb Wing was moved here in April 1944 to give their planes solid striking power at targets farther north. These all were tactical bomber groups whose focuses were Axis positions, resources, and supply lines across those pristine waters in Italy. It was here that all of the action of *Catch-22 could* take place and here that all the factual action of Heller's Bomb Group *did* take place.

George's memories are keen as he relates:

Other than placing the Group on that small island, I find the setting very factual. All places mentioned—countries, islands, towns, cities, etc.—are actual places and they did play a part in our group's history. I found no place mentioned that did not. Even the location of targets and the types of targets for those locations are both correct for targets hit by the Group. There is no doubt in my mind that Heller took actual events and actual people to write his story. I think he varied the people used for a particular event on occasion, as well as split some people into two different characters. He uses names at certain times, which are very close to a real person who actually did the thing mentioned. He has also incorporated some events, which took place before he arrived but that he became aware of by word of mouth.

The Italian island of Pianosa and the French island of Corsica.

George had become a seasoned warrior by the time he encountered Joe Heller, who was a combat replacement, fresh, eager and, yes, naïve. George, 5'7" and just 24 years old, flew his first combat mission on October 26, 1943, mere months before 5'11½ " and just 21-year-old Lt. Heller arrived in the Group. But those few months were a lifetime in terms of experience. While each barely knew the other, still they shared, for a momentous period of time, a history together. They shared the adjustment of going from a life of the familiar—family, friends, ball games, the life of the American boy on American turf—to the unfamiliar of guns, war planes, death, and responsibilities not for your friend's missed homework assignment, but for your friend's very fragile life in a very foreign place.

As a pilot, George was keenly aware that he shouldered the lion's share of responsibility for the safety of his entire crew. It was for him to decide the critical timing of almost every movement of his aircraft which affected, as well, the following five formation-flying B-25s in his box of six. It was for him to keep vigilant, use his every left-seat skill to evade gunfire and flak, and demand of his aircraft instant evasive maneuvers after "Bombs Away!" His was the decision, when his plane was riddled and crippled from the accuracy of the anti-aircraft guns manned by the tense and sweating hands of the German foe so far below them, whether his men should bail out with all of the risks that entailed or whether he should risk their lives further by struggling to bring his mortally wounded plane safely home.

Joe also wore a heavy cloak of responsibility. As a bombardier he learned to steel his nerves. This was not a position for the timid. As he sat in his exposed Plexiglas nose cone, in front of, and isolated from, the rest of the crew, he had to focus, with a singularity, on one paramount pinpoint target. When the B-25 made its final turn and veered toward this target, it was Joe, as bombardier, who took complete control and called the shots. Nothing was to distract from his tunnel vision of guiding that aircraft over the target. His bible was the continually drifting cross hairs on his gauge. His sole focus was to accurately drop his bombs on a tiny speck of an objective some hundreds of miles away. If shells were exploding around him and death-dealing flak was ripping through his aircraft's skin, he must not be distracted. The entire crew knew that should they—should Joe, in fact —miss their target, they would be assigned to return on the following day to complete it, this time under even more dire conditions, since the enemy below had been alerted and would be waiting. It was every man's most urgent prayer that the bombs would drop with accuracy and their aircraft could instantly bounce up free of their weight, bank, climb to safety . . . and head home. As Joe glued himself to his bombsight, their lives were put on hold.

Joe Heller's and George Well's wartime experience centered on their crewmates, their leaders, their aircraft, and their surroundings. The island of Corsica was the hub of their universe. Their group was comprised of four squadrons: the 486th, 487th, 488th, and 489th. Heller, Wells, and most, but not all, of the men who surface in some form in *Catch-22* were part of the 488th. To better understand the genesis of *Catch-22*, a very brief history lesson of the 340th Bomb Group and its aircraft is needed.

As the warhorse Bucephalus was to the great Alexander, so the beloved B-25 Mitchell medium bomber was to the 340th Bomb Group. It was the protector of every crewmember who boarded her. The crew bonded tightly with her. They named her *Oh! Daddy!*, *Briefing Time*, *Poopsie*, *Black Jack*, *Jersey Bounce*, *Kick Their Axis*, *Vesuvianna*, *That's All-Brother*, *Battlin' Betty*, etc. and credited her every accomplishment with small bombs (or in *Poopsie*'s case, small puppies) painted, like Wild West notches on a pistol butt, on her fuselage. They loved and honored her, for she could sustain tremendous damage and still be able to limp home to safety with her cherished crew.

This B-25 was the ideal aircraft for medium altitude daytime attacks on specific targets. The high percentage of pinpoint bombing accuracy maintained by the 340th, 310th, 319th, and 321st Bomb Groups resulted in flying sorties on precisely chosen tactical targets that the higher flying heavy bombardment groups had not succeeded in knocking out. The average lower bombing altitude of from 7,000 to 12,000 feet insured a markedly increased degree of accuracy, but also brought the B-25 into the deadly range of the highly respected and effective German 88mm cannon.

This high-velocity gun was Germany's main defensive weapon against the bombers. The 8.8cm *Fliegerabwehrkanone*, shortened to Flak, fired a 20.34-pound shell to over 49,000 feet. The weapon was evolved to combat the fact that to hit a moving aircraft flying about 200 MPH at an altitude of one or two miles with a shell was exceedingly difficult. This weapon was designed to fire ahead of the formation and was timed to explode when the planes reached that spot. The explosion would rocket hundreds of slashing, iron shards in all directions. Since the artillery man would have to guess at the altitude of the formation, the cannon shells contained a timer that would allow him to set them to go off at, for example 5,000 feet, or 5,500 feet, and so on. That is why pilots originally employed severe evasive maneuvers, for it kept those feared gunners below guessing as to where to aim their next volley. True to life, Murphy's 10th Military Law states, "Never worry about a bullet with your name on it. Instead worry about shrapnel addressed, 'Occupant'."

Heller's descriptions in *Catch-22* of the wildly erratic evasive flying maneuvers of Yossarian were wholly accurate in the initial bomb runs. George's personal descriptions attest to that. Knowing they had a deadly gauntlet to run, he had to keep razor alert for the upward snake of smoke trails signifying incoming ack-ack, that dreaded anti-aircraft gunfire. Instantly reacting to this sight, he would roll his aircraft first to the right, then left, up, and then down until within about one-half mile of the target. Then for the next five to ten minutes he would fly unwaveringly *straight*, his men in gripping focus, tense, alert, with pulses racing and breathing halted, to hit the target. This was the most dangerous time. The bombs would drop and immediately the planes left the target. The remaining five B-25s in this box of six aircraft would be following suit.

While the 340th had a wide variety of targets (airfields, railroads, bridges, road junctions, supply depots, gun emplacements, troop concentrations, marshalling yards, and factories) it was the Brenner Pass, a narrow mountain pass through the magnificent jutting Alps along the border between Italy and Austria, that was their greatest challenge.

Historically this pass was always of strategic importance. At 4,495 feet it was the lowest of the major alpine passes, as well as one of the few in the area. During World War II the Alps impeded the German infantry that was being pushed by the Allies, from reaching the safety of Austria. The only way to cross the Alps was through this mountain pass at Brenner. As well, the pass was the route used by the Germans to funnel supplies down from Austria to their ground troops in Italy. Because of the importance of this pass, the Germans kept it heavily armed with artillery of all kinds.

In November of 1944 the medium bombers were given their most daunting test, one they accomplished with amazing success. It was known as the Battle of the Brenner, an intense focus on this critical rail line between Germany and the Italian battlefront. Targets, ranging into Austria, were the railroad bridges, tunnel entrances, or replacement troop or supply concentrations.

Typically an alpine mission would unfold as follows. The designated target might be the marshalling yards outside the city of Trento. The attack was planned so the bomb drop could be made at high noon. This timing was essential since, for visual bombing, the target must be well seen, and only during the noon hour does the sun shine on the floor of the Brenner Pass. The surrounding high mountains cast their deep, dark shadows during any other period of the day.

The sortie would be accomplished at an altitude of from 100 to 300 feet above the mountain peaks, coming in over the pass, turning immediately on the IP,[1] and making the bomb run up the Brenner. Naturally the Axis forces soon recognized this and prepared their defenses along those lines—even to the extent, as it was later learned, of scheduling earlier lunch hours. Obscuring smudge pots were kept burning from 1100 hours each day in strategic areas.   In addition, the fierce 88mm anti-aircraft guns were transferred from the floor of the valley to positions high on the mountainside where they could actually fire down upon B-25s that swooped through the pass at low level.

MISSION #95: DEC. 22, 1944
Formation Commander with 489th on target at Lavis on the Brenner Pass line. Ran into lots of ack-ack and picked up quite a few holes. Really was a cold ride for four and a half hours.

On rare occasions the medium bomber groups were afforded the luxury of anti-flak fighters assigned to attack these enemy gun positions, thereby greatly increasing the odds of US bombing success and safety.

The increasing accuracy of German gunners also led the medium groups to devise their own means of defense against the anti-aircraft fire.   For example, the creative use of chaff, strips of metal foil released in the atmosphere from aircraft to confuse radar-tracking missiles or obstruct radar detection, was implemented.   Eventually, three to six B-25s were loaded with *verboten* phosphorous bombs in advance of the main bomb group, with the enemy gun positions as their targets. This ever-increasing determination and creativity reaped rewards.

The B-25s operated during the entire winter months of 1944 from the island of Corsica. This winter was not to be taken lightly at altitude. Mother Nature may have railed against man's weakness for wars by trying to clear her skies with icy, finger-freezing, mind-numbing cold, to which George's 98th mission attests. (*see page 30*)

The B-25s had no heaters so, in the manner of iconic comic illustrator Charles Schultz's *Charlie Brown*, men layered heavily before venturing up. Had Hitler read and heeded *Der Peanuts*, perhaps Moscow, in those earlier days, would have been rechristened Hitlersburg.

Victor J. Hancock, of the 445th squadron, describes, in the winter 2010 issue of *Men of the 57th*, the crew's *couture*:

First on were your boxers followed by wool long johns. Over those would be the wool trousers, wool shirt and wool sweater. Pull a wool beanie over your head. Silk socks, regular socks, then wool socks. On with fleece-lined pants. Placed over those were combat boots that were placed inside fleece-lined boots. Then they pulled on a flight jacket, over that a heavy leather fleece-lined jacket and on their head a fleece-lined helmet. Hands were covered with silk gloves, over which they placed leather gloves over which they then placed fleece-lined leather gloves.

Now just try to pull out a tissue.

From this Corsican island, in the months of summer, the assault that opened the front in southern France was conducted.

MISSION #68: JUNE 29TH

Flew Formation Commander with 488th Sqd. Target railroad bridge at Cervo, Italy. I got my first look at French soil today because the target was very close to France.

It looks as though we'll soon be working on Southern France.

On D-Day for Operation Dragoon, the invasion of southern France, it was the B-25 units that were assigned the low-level attack on enemy concentrations. In this type of operation it was not an unusual occurrence for bombs from the higher-flying heavy US bombers to rain down on the B-25 formation. (As George ducked on Valentine's Day, February 14, 1944, "We nearly had bomb hit us.")

The B-25 group's four squadrons each maintained approximately fifteen assigned aircraft. A normal mission involving the entire group consisted of three participating squadrons, each squadron providing twelve aircraft. The standard formation was boxes of six aircraft in two, tight V formations, the back three dropping a plane's depth below the fore three for visual purposes. Grouping aircraft in this manner provided the most effective defense against enemy fighters by the concentration of firepower.

This same box formation was held throughout the bombing run, five of the aircraft dropping off the lead ship. The lead aircraft in each box was generally the only aircraft containing a bombsight; release of the bombs in the remaining five aircraft was either by visual toggling or radio signal release. The usual armament of a B-25J was four fixed .50 caliber machine guns firing forward, one flexible .50 in the nose, twin .50s in the top turret, two .50s in the waist, and twin .50s in the tail, the stinger.

Each ship carried a crew that consisted of a pilot, copilot, bombardier, turret gunner, radio operator/waist gunner, and tail gunner. The dominant lead ship in each formation bore, as an additional crewmember, a navigator. It was the last ship that generally also carried an additional crewmember, a photographer to document the mission's success or failure.

The pilot, copilot, and bombardier/navigator were officers, while the gunners were enlisted men; each was a critical piece of the puzzle. From front to back, first came the bombardier snugged in the nose of the plane. Above and behind the nose sat the pilot and copilot who, while flying the plane, were continually straining to avoid barrages of flak, simultaneously firing the forward facing guns. The navigator, usually behind the pilot and copilot, was in constant communication with the bombardier. The navigator's job was to guide the plane toward the target, while the bombardier had to release the bombs with flawless timing for a successful mission. In the body of the plane were the bomb bay and the radio compartment. Stationed here were the radio operator and engineer, who both did dual duty as waist gunner and turret gunner. Farther back rode the more isolated tail gunner. At takeoff this man would not be in the tail gun turret, but instead would be closer to the body of the plane so that his extra weight would not affect the B-25's sensitive aerodynamics as it lifted off the runway.

On Sunday, January 21, 1945, an accident happened in the Brenner Pass. Because of brutal air turbulence, B-25 formations on a mission were bouncing around violently with great fluctuations in altitude by each of the planes. Two runs on the target and they still could not drop. On the third run, battling both heavy flak and fierce air gusts, the propeller of aircraft 8U, piloted by 1st Lt. William Y. Simpson, struck the tail gunner's compartment of aircraft 8P containing S/Sgt. Aubrey B. Porter. Porter was actually cut out of the tail by the prop.

Radio-gunner Jerry Rosenthal, of the 488th, gave the following account: "Porter fell through space without a chute. His chute and part of the 8U's wing came by our tail. The whole thing just missed us . . . " Simpson's plane went down over the target. The pilots of the severely crippled 8P pulled an "aerodynamic miracle" when they eventually were able to safely land their B-25 minus more than half of its tail. And, tragically, its tail gunner. Aubrey B. Porter was never found.

B-25, 8P, missing half of its tail and its tail gunner.

In *Catch-22*, Heller describes this collision between the two aircraft on a mission. Young pilot, William Simpson, is immortalized as "Kid Sampson."

Following is a flight schedule from this same 488th Bomb Squadron for an earlier mission on August 23, 1944 in which the same two aircraft, 8U and 8P, were participating. 8P had 2nd Lt. Joseph Heller as bombardier and 8U carried 2nd Lt. Frances Yohannan, also bombardier, whose name was the inspiration for the name of *Catch-22*'s main character, Yossarian. Aircraft 8U and 8P were flying adjacent to each other in the box just as they were on that fateful Sunday. Only fate decided which day, which people.

```
                            488TH BOMBARDMENT SQUADRON (M) AAF
                             340TH BOMBARDMENT GROUP (M) AAF

                                                    APO 650, c/o Postmaster,
                                                    New York, N. Y.,
                                                    23 August, 1944.

        SUBJECT:  Ships and Crews to fly in second combat mission, this date.

        TO     :  Operations Officer, 340th Bombardment Group (M) AAF.

             1.  Briefing at 1520 hrs.  Pre-briefing at 1500 hrs.
```

			NO. 43-27669 (8T)				
			1st Lt. S.Aswad	P			
			Capt. H.B.Howard	CP			
			1st Lt. W.A.Davidson	N			
			1st Lt. R.R.Burger	B			
NO. 43-27695 (8J)			T/Sgt. J.Lazor	RG		NO. 43-4055 (8M)	
1st Lt. J.B.Rome	P		S/Sgt. C.H.Snow	G		2nd Lt. R.R.Shipman	
2nd Lt. J.D.Kroening	CP		T/Sgt. W.H.Hudgins	TG		2nd Lt. H.L.Lacey	
2nd Lt. N.S.Rosenthal	B					2nd Lt. H.A.Moody	
Cpl. C.S.Squires	RG		NO. 43-35983 (8E)			T/Sgt. W.P.Day	
Sgt. S.Woytek	G		1st Lt. H.C.Gross	P		S/Sgt. H.E.Bartell	
S/Sgt. E.L.Headley	TG		2nd Lt. B.D.King	CP		Sgt. S.Slimowitz	
A.J.Lockhart			2nd Lt. W.O.Fischer	B			
NO. 43-27504 (8K)			T/Sgt. B.Greenbaum	RG		NO. 43-3990 (8B)	
2nd Lt. C.C.Grosskopf	P		S/Sgt. V.A.Gasperino	G		2nd Lt. G.J.Hellyar	
2nd Lt. E.M.Holtz	CP		S/Sgt. A.E.Rosin	TG		2nd Lt. E.J.Kirk	
2nd Lt. F.A.Pfeffer	B					2nd Lt. H.F.Robinson	
Cpl. W.F.Ziegler	RG		NO. 43-27752 (8N)			Cpl. W.E.Pond	
S/Sgt. A.E.Ewan	G		Capt. J.E.Rapp	P		S/Sgt. E.A.Iljana	
Pvt. C.M.Ricks	TG		1st Lt. B.A.Steed	CP		Sgt. M.R.Tafoya	
			Capt. C.T.O'Brien	N			
NO. 43-27657 (8P)			1st Lt. T.C.Sloan	B		NO. 43-4025 (8Q)	
1st Lt. W.B.Reagan	P		T/Sgt. L.E.Anderson	RG		1st Lt. J.J.Swift	
1st Lt. E.G.Sallen	CP		S/Sgt. B.E.Gorski	G		2nd Lt. E.J.Ritter	
2nd Lt. J.Heller	B		T/Sgt. T.L.Higgins	TG		2nd Lt. D.L.Atkinson	
T/Sgt. F.Mirochnick	RG					T/Sgt. W.S.Goodell	
S/Sgt. H.L.Rackmyer	G		NO. 43-27474 (8R)			S/Sgt. W.E.Porter	
S/Sgt. J.E.Smith	TG		1st Lt. E.W.McDonald	P		S/Sgt. R.A.Kerkhan	
			2nd Lt. R.R.Vertrees	CP			
NO. 43-4064 (8U)			2nd Lt. E.E.Bardnell	B		NO. 43-27537 (8Z)	
1st Lt. G.W.Clifford	P		T/Sgt. R.J.Martin	RG		1st Lt. M.G.Duncan	
2nd Lt. A.A.Gmachl	CP		S/Sgt. R.Bland	G		1st Lt. J.F.Mummey	
2nd Lt. F.Yohannan	B		S/Sgt. J.W.Ryba	TG		2nd Lt. A.S Householder	
T/Sgt. A.L.Green	RG					T/Sgt. J.A.McGloin	
S/Sgt. N.E.Klinkner	G					S/Sgt. R.D.Chesney	
Cpl. P. Sims	TG					Sgt. A.J.Bertagna	

```
                                            JOHN E. RAPP,
                                            Capt., Air Corps,
                                            Operations Officer.
```

Combat mission includes Heller and Yohannon.

As George's crew lifted off for yet another mission, they knew the routine. This mission, from Corsica to the Brenner Pass, was to last approximately 3½ hours. Their aircraft labored with its bomb load of about 5,000 lbs. They threaded their way around known anti-aircraft gun positions trying to avert the *piñata*-like attacks of those fearful weapons. The sounds had become unwillingly familiar. The staccato ack-ack fire from ground installations was of two types over this target: barrage coming in, and tracking going out. A sudden stop in firing indicated enemy fighters closing in[2] and every man aboard would suck in his breath as he frantically scanned the sky for sight of the lethal and rapidly approaching, but still faceless, enemy.

Upon his return to home base, George would pull out his small, black mission book and dependably log each of his 102 missions. This soft and functional diary with white, lined pages measured about 3x4 inches and easily slipped into his shirt pocket. The entries were brief and devoid of emotion. *"Mission #57: May 16, 1944. Led my Sqd. Flight on Port of Piombino. Lots of flak but we dove around it and only got two holes."* These daily-abbreviated entries fleshed out the true story of a pilot's circumstances. Entries would show, curiously, how sometimes the most difficult missions were the least storied.

> *98th Mission—Jan. 28th.   Flight leader with 488th on R/R Bridge at Roverto, Italy lots of flak and this was the hardest mission I've had to date due to cold and no oxygen."*

Mission entries record how he flew even when he was very sick from the liver ailment, yellow jaundice

MISSION #14: DEC. 5TH

> *Was very sick but flew anyway. Co-pilot for Dean on raid to Aguilla. Bad weather made us bring our bombs back but ran into a lot of ack-ack on coast. Couldn't get left engine out of high blower.*

Faithfully, he penned in his mission entry. He flew, as well, when his plane was very sick, returning from one mission with 33 holes in it.

MISSION #20: JAN. 14TH

> *Flew as leader of 2nd element on Pontercorvo Bridge, Italy. My plane had 15 holes in it, seven of which were in the right engine nacelle (cover). The left engine was hit and leaking oil. The right main gear tire had been hit and blown out leaving me with no tire or brakes for the right wheel. Other hits in wing and bombarding compartment. I made a good approach to landing but after I touched the ground, I had my hands full. When I finally got the plane stopped, we were facing the opposite direction! The group lost 2 ships on raid.*
>
> *I've been sick with yellow jaundice for the past week. Just found out that my plane had 33 holes in the raid on the 14th. The hydraulic system was shot out in the right nacelle*

Again, penning his entry. He flew under conditions that left every plane in his flight suffering from direct hits. The entry was recorded. He flew sick or well, in good weather or bad, occasionally returning from one mission then taking off for another, and occasionally a third that same day. Settled in his tent in the quieting evening hours, he would routinely thumb open his book and, in that very meager space, record those missions.

MISSION #9: NOV. 29TH

Flew copilot for E.J. Smith. Supposed to bomb Terni but weather kept us from seeing the target. Then we flew to east Coast, we bombed Giulianova, Italy. Bombed bridges and marshalling yards. Ran into a lot of flak as we came off the target. Must have been 105mm because they didn't fire straight up. 33 B-25s unescorted.

MISSION #10: DEC. 1, 1943—WITH R.M. JOHNSTON

Enemy positions near Casino, Italy. Took off in morning but called back on account of weather. Take off again in afternoon. Johnston couldn't stay in formation and went over target by ourselves! 5,000 lbs of bombs. Did not get credit for mission because bombs were not dropped. Johnston had to go around on coming in for landing. Later got credit. Dec. 3rd.

George, of humble beginnings and unsurpassed optimism, had become the ideal pilot. While his smaller, strong frame fit well in the confines of the B-25, and his talent was patent, it was his mental outlook that was invaluable. Never did he doubt he would return home once the war ended and that unshakable belief blessed him that most rare of gifts: freedom from fear.

He relates how:

We [he and Fred Dyer, *Catch-22*'s Captains Piltchard and Wren] both did like to fly a lot and we became better by flying more. Dyer and I would go up and fly on each other's wing many times on stand-down days or when we weren't on a mission. We would fly in heavy weather conditions in formation—practice single-engine in formation. We flew the planes in practice under various conditions to fully understand its performance as well as its limitations.

MISSION #101: MARCH 13, 1945

Formation Commander with 488th on Aldeno R/R. Fill in the Brenner Line. Had lots of flak and had an oil line hit in the right engine and had to go on single engine over the target. We had to drop out of formation but we managed to get back OK. This was the 1st medium bomber that has ever returned from the Brenner Line on single engine. We were on single engine for 2 hours. We were able to hold it at around 6,500 ft. We were shot at again crossing the Po Valley by 40 and 20 mm. We then had a hard time getting over the mountains between Po Valley and the coast.

We really set the tone for combat in the 340th. What we both did was to tend to eliminate the basis for bellyaching by the combat crews due to any concern over the B-25's performance even when heavily damaged or due to fear for one's self while in combat. We flew additional missions without having to, never picking one because it was considered a milk run. In fact, we did the opposite. The more challenging, the more we competed with each other. We were the best of friends. I used to call him "Fearless 'Freddie' Fosdick" after the Dick Tracy comic strip character.

Fred's love of flying combined with a curious and adventurous nature to produce a story—a very short story.

While I was stationed at Comiso, Sicily, a pile of wrecked remains of over one hundred Me-109s was discovered. One had been flown by the German ace Molders. His plane had over twenty swastikas painted on the fuselage (Note: Werner Molders was the first pilot in aviation history to claim 100 aerial victories. It paid to be born "Allied" rather than "Axis." The Luftwaffe, during WWII, far outstripped allied fighters in numbers of recorded kills because of their "fly till you die" policy rather than being rotated elsewhere after a certain number of missions had been completed.) I and two others put a couple of these airplanes back together and flew them. The Me-109 had a narrow landing gear, which was (according to German sources) notoriously weak. The second time I flew mine, all did not go well. The fact that the oil pressure was calibrated in kilograms/cm. instead of lbs./in. was confusing. On attempting to land, one of the main gears wouldn't go down so, after flying around trying to shake it loose, I eventually landed on one gear and a wing amid a great cloud of dust.

We eventually rebuilt the plane and, on the move to Catnip, Sicily, I asked an ex-RAF pilot to fly it to Catania for me. On take-off, he forgot to latch the canopy, which blew open; he ground-looped the plane, wiped out the landing gear, and that was the end of that.[3]

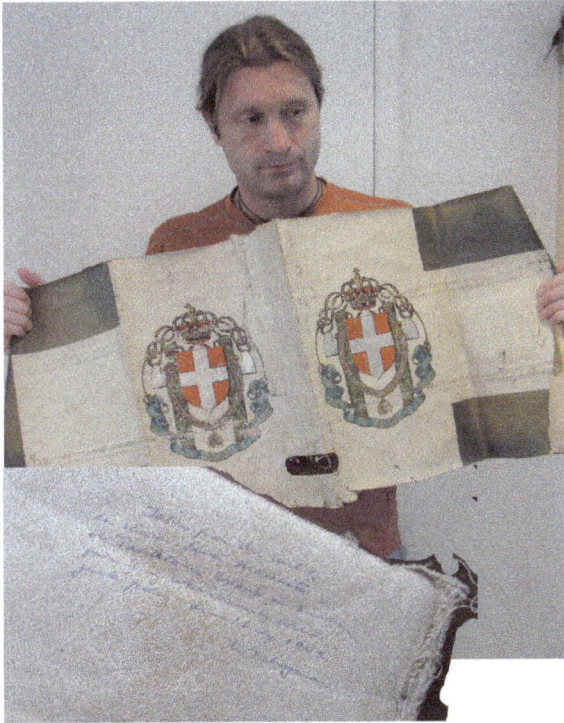

Col. Chapman's grandson, Jason, holding skin from 1943 wrecked Italian aircraft.

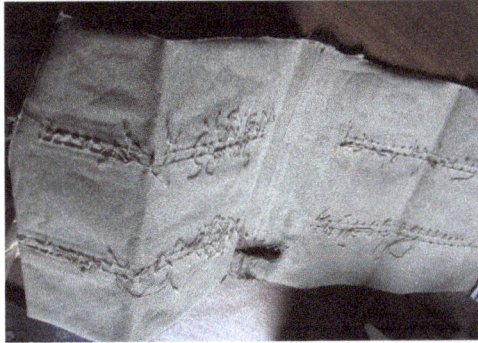

Coarsely stitched underside of Italian aircraft fabric skin. In the upper corner Bill had written in ink: "Taken from the tail of an Italian Siai Marchetti on Pantilaria Island in Mediterranean from the war casualty wrecks junk pile. June/ July 1943
—W.F. Chapman"

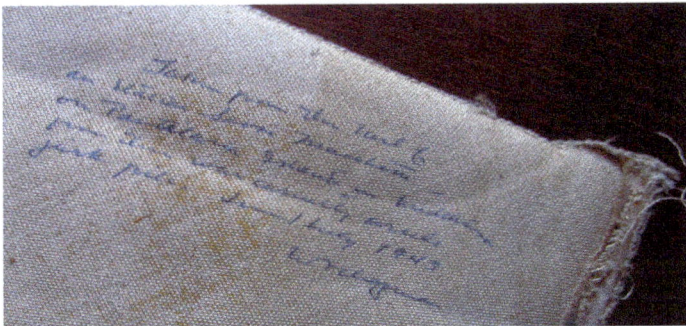

Pure and simple, both men loved to fly. And Joe, on the sidelines, nimbly massaged George and Fred into *Catch-22*'s very identifiable, quiet and capable Piltchard and Wren. Both men felt, as they trained younger pilots, that men who tell other men how to fight should fight alongside them, as well.

After his historic 102nd mission, Fred, pilot extraordinaire and war-ripened at age 29, was firmly ordered to return home. He had survived three engine losses in combat, uncountable ack-ack holes, a shot-out windshield that resulted in a head wound, and, ultimately, had been shot down, parachuting from a fire-encased aircraft only after ensuring all crew members had safely donned parachutes and jumped first. Bad fortune dropped him in the midst of a tank battle, from which good fortune allowed a rescue by British soldiers.

Fred reflects:

> Every mission had its distinct personality or ambiance, so to speak. I think that one has to admit that all were fraught with trepidation. Some had their moments, though; like the time we were returning from a night mission over Battipaglia, Italy; the moon was shining and we could see Stromboli down in the dark waters of the Mediterranean. If German night fighters didn't show up, you had it made. At other times when a mission returned to home base, the formation went into an echelon for landing. Each airplane peeled off at three-second intervals. If you made it this far, the rest was easy; this was a great moment

Return home he did, probably with an audible sigh, for it was with great reluctance. "I just hate to see this show going on without me in it."
Go home, they said. That's it. That's enough.
George, too, was ordered home, but his leave was for a 30-day rest.

MISSION #102: COMMAND PILOT WITH 489ᵀᴴ ON MARCH 19ᵀᴴ, 1945

This was the Group's 800th mission. The target was a R/R Bridge at Muhldorf, Austria. This should be the last mission I'll fly before going home for a thirty-day leave.

AFTER THE WAR

Frederick Wolfin Dyer, Jr., with a chest full of medals attesting to his bravery, skill, and experiences, remained in the Air Force, rising to the rank of colonel while continuously flying all manner of aircraft. This native Coloradan retired after thirty years' service, having accumulated about 10,000 flying hours. He bought, operated, and eventually sold a successful transmission repair franchise with his eldest son. Then he moved on to the machine design and antique car restoration business. At age 80, three years before he died, Fred and one of his sons climbed Mount Revenue, a 13,000 foot peak twelve miles east of Montezuma. "This guy did not do Senior Citizen," his son said. "He just never got there."

The Box of Six: standard flying formation.

CHAPTER 2

JOE: WHO SHARES THE SAME DNA AS **YOSSARIAN**

Joseph Heller had a pre-war life. He was a son of Brooklyn. He was also the son of two first-generation Russian-Jewish immigrant parents, Lena and Isaac Heller. When young Joey was a malleable and impressionable five year old, his father, a wholesale bakery truck driver, died after a bungled surgical operation. Joey, his mother, and his two devoted older siblings, who he discovered as an adult were half siblings from a first marriage of his father, were left to survive as best as they could in the carnival-like atmosphere of Coney Island. This included his mother accepting boarders into their four-room apartment. Such an unusual environment, besides giving him a pure, unreconstructed, streets-of-New York accent—a kind of high nasal blat—is credited with encouraging his ironic, smart aleck nature and wry humor, the supporting structure of his literary style. His sometimes-perplexed mother was known to say, "Joey, you got a twisted brain."

In 1941 he graduated from Abraham Lincoln High School, working lightly delivering telegrams. A life turning, "ah-ha" moment had occurred after reading a child's version of Homer's Iliad that convinced him, if he ever grew up, that he wanted to be a writer. His first job, however, that of becoming a blacksmith's helper, was not exactly the first step.

After the United States entered World War II, sharp-witted Joe and a few Brooklyn buddies, full of life and bursting with patriotism, enlisted in the Army Air Corps and, with much grandstanding, boldly took the oath of enlistment in Grand Central Terminal. He remembered his mother weeping as the trolley car pulled away with him in it. He couldn't figure out why she was so unhappy. He felt like he was going to Hollywood.

War seemed quite exotic for Joe; the standard of living was higher in the Army than in Coney Island, he ate better, and he had more money in his pocket than ever before. And the Army offered travel.

So he traveled to armorer's school. He would have been quite content to remain there; people who made armor and affixed it to war machines did not usually go into combat. He heard, however, an erroneous rumor that armorers were being turned into gunners and gunners' lives were "worth no more than three days." Off he went to cadet school and became a bombardier. He was commissioned a 2nd Lieutenant and traveled some more.

After crossing the Atlantic Ocean, Joe stepped onto Corsican soil and his life began a very slow creep from the levity of a youth on a glamorous Jimmy Stewart/Clark Gable type of adventure to, eventually, that of a trained and skilled warrior whose life was up for grabs on each mission.

The war eventually ended, many years passed, and one evening, while lying in bed thinking other thoughts, Joe surprised himself with the opening lines of the yet unconsidered *Catch-22*. The next morning he wrote the entire first chapter and sent it to his agent, who sold it to New World Writing. He recalls how "I was so excited I couldn't wait to begin chapter two. One year later, I did."

Fast forward further to 1961. Judgment Day surfaced. With the eventual publication of his first novel, *Catch-22*, sides were chosen. The readers divided.

As the book gained in popularity there was one side that hailed 39-year-old Heller as their tongue-in-cheek voice to deride and protest. The Vietnam War fanned these flames. On the other side were those who railed against the author's use of sarcasm, black humor, and ridicule against authority, the military, and his war mates. The temptation was to question Heller's wartime abilities and loyalty.

But take another look. Heller was a bombardier. Consider the bombardier.

To reach the bombardier's station, that "greenhouse" in the nose of the B-25, the bombardier, as opposed to every other crewmember, had to leave his parachute at the entrance of the short, claustrophobic tunnel through which he must crawl. The farther away from the pilot that a crewmember was, especially the isolated bombardier and the backward-facing tail gunner, the more fortitude it took to overcome apprehension from lack of instant information as to what was happening in the rest of the plane, not knowing how drastic the situation was, and the awful feeling of being alone in critical circumstances.

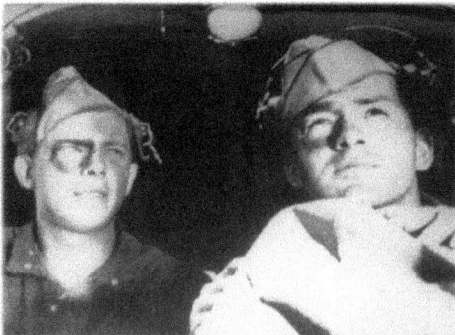

"The Badge of the Bombardier," a black ring around his right eye from the bombardier's rubber eyepiece.

The bombardier sat out in front with his Plexiglas panoramic view of the approach to the target and its defenses. Once the battle began, he was put right in the middle of it. Under certain conditions from this remote position it was impossible for the bombardier to bail out if it became necessary. It took an exceptionally brave man to ignore the battle and concentrate to the degree needed to take out a pinpoint target such as a twenty-foot bridge from over two miles away. Some could do it; others could not.

Joseph Heller flew 60 missions.

> The crawlway was Yossarian's lifeline to outside from a plane about to fall, but Yossarian swore at it with seething antagonism, reviled it as an obstacle put there by providence as part of the plot that would destroy him. There was room for an additional escape hatch right there in the nose of a B-25 but there was no escape hatch. Instead there was the crawlway, and since the mess on the mission over Avignon he had learned to detest every mammoth inch of it, for it slung him seconds and seconds away from his parachute, which was too bulky to be taken up front with him, and seconds and seconds more after that away from the escape hatch on the floor between the rear of the elevated flight deck and the feet of the faceless top turret gunner mounted high above. (p.48)

With a stew-pot of war-determined circumstances and a sea of both military and civilian personalities surrounding him, Joseph Heller had little need to look farther than arm's length for literary material. Scenarios such as the following filter through *Catch-22*, providing solid confirmation and example of how his facts transitioned into fiction. Couple these personal experiences with his fertile imagination and out pops—Luciana!

> Luciana fled mirthfully along the sidewalk in her high white wedgies, pulling Yossarian along in tow with the same lusty and ingenuous zeal she had displayed in the dance hall the night before and at every moment since. Yossarian caught up and walked with his arm around her waist until they came to the corner and she stepped away from him. She straightened her hair in a mirror from her pocketbook and put lipstick on.
>
> "Why don't you ask me to let you write my name and address on a piece of paper so that you will be able to find me again when you come to Rome?" she suggested.
>
> "Why don't you let me write your name and address down on a piece of paper?" he agreed.
>
> "Why?" she demanded belligerently, her mouth curling suddenly into a vehement sneer and her eyes flashing with anger. "So you can tear it up into little pieces as soon as I leave?"
>
> "Who's going to tear it up?" Yossarian protested in confusion. "What the hell are you talking about?"
>
> "You will," she insisted. "You'll tear it up into little pieces the minute I'm gone and go walking away like a big shot because a tall, young, beautiful girl like me, Luciana, let you sleep with her and did not ask you for money."
>
> "Stupido!" she shouted with emotion. "I am not asking you for any money!" She stamped her foot and raised her arm in a turbulent gesture

that made Yossarian fear she was going to crack him in the face again with her great pocketbook. Instead, she scribbled her name and address on a slip of paper and thrust it at him. "Here," she taunted him sardonically, biting on her lip to still a delicate tremor. "Don't forget. Don't forget to tear it into tiny pieces as soon as I am gone."

Then she smiled at him serenely, squeezed his hand and, with a whispered regretful "Addio," pressed herself against him for a moment and then straightened and walked away with unconscious dignity and grace.

The minute she was gone, Yossarian tore the slip of paper up and walked away in the other direction, feeling very much like a big shot because a beautiful young girl like Luciana had slept with him and did not ask for money.

[Later at the dining table] Yossarian choked on his toast and eggs at the enormity of his error in tearing her long, lithe, nude, young vibrant limbs into tiny pieces of paper so impudently and dumping her down so smugly into the gutter from the curb. He missed her terribly already. *(pp.161-162)*

About this account, Heller confesses:

His [Yossarian, *Catch-22*'s bombardier hero] encounter with Luciana, the Roman whore, corresponds exactly with an experience I had. He sleeps with her; she refuses money and suggests that he keep her address on a slip of paper. When he agrees, she sneers, "Why? So you can tear it up?" He says of course he won't and tears it up the minute she's gone —then regrets it bitterly. That's just what happened to me in Rome. Luciana was Yossarian's vision of a perfect relationship. That's why he saw her only once, and perhaps that's why I saw her only once. If he examined perfection too closely, imperfections would show up."[1]

The imagination is only mildly stretched by watching Joe slip into Yossarian. As surely as mild-mannered Major Joe Ruebel evolved into equally affable Maj. Danby or the more mature Major Jerre Cover stepped into grey-haired Major _____de Coverly, so did Yossarian develop from the Essence-of-Joe. Like a sculptor, Joe manipulated his media—add here, remove there, enhance, exaggerate—but the supporting structure remained, yes, Joe. Yossarian safely allowed Joe Heller to be Joe Heller.

With the keen anticipation of the inexperienced, Joe was introduced to his first mission four days after his arrival in Corsica.

The first time I came to Corsica was in May, 1944, when I joined the bomb group as a combat replacement. After four days I was assigned to my first mission as a wing bombardier. The target was the railroad bridge at Poggibonsi.

Poor little Poggibonsi. Its only crime was that it happened to lie outside Florence along one of the few passageways running south through the Apennine Mountains to Rome, which was still held by the Germans. And because of this small circumstance, I had been brought all the way across the ocean to help kill its railroad bridge.

The mission to Poggibonsi was described to us in the briefing room as a milk run—that is, a mission on which we were not likely to encounter flak or enemy planes. I was not pleased to hear this. I wanted action, not security. I wanted a sky full of dogfights, daredevils and billowing parachutes. I was twenty-one years old. I was dumb. I tried to console myself with the hope that someone, somewhere along the way, would have the good grace to open fire at us. No one did.

As a wing bombardier, my job was to keep my eyes on the first plane in our formation, which contained the lead bombardier. When I saw his bomb-bay doors open, I was to open mine. The instant I saw his bombs begin to fall, I would press a button to release my own. It was as simple as that—or should have been.

I guess I got bored. Since there was no flak at Poggibonsi, the lead bombardier opened his bomb-bay doors early and took a long, steady approach. A lot of time seemed to pass. I looked down to see how far we were from the target. When I looked back up, the bombs from the other planes were already falling. I froze with alarm for another second or two. Then I squeezed my button. I closed the bomb-bay doors and bent forward to see where the bombs would strike, pleading silently for the laws of gravitational acceleration to relax just enough to allow my bombs to catch up with the others.

The bombs from the other planes fell in an accurate, concentrated pattern that blasted a wide hole in the bridge. The bombs from my plane blasted a hole in the mountains several miles beyond.

It was my naïve hope that no one would notice my misdemeanor; but in the truck taking us from the planes a guy in a parachute harness demanded, "Who was the bombardier in the number two plane?"

"I was," I answered sheepishly.

"You dropped late," he told me, as though it could have escaped my attention. "But we hit the bridge."

Yeah, I thought, but I hit the mountain.

As noted, in aircraft 8A flew Joseph Heller as bombardier and Lt. Joseph Chrenko as pilot (Chrenko was Yohannon's tent mate and the basis for the character Hungry Joe in *Catch-22*. In aircraft 8F flew George Wells as lead pilot.

Twenty-plus years later, Heller revisited Corsica and Italy and noted:

A few days after I returned to Italy from Corsica with my family, we rode through Poggibonsi on our way south to Siena to see an event there called the Palio. The railroad bridge at Poggibonsi had been repaired and is now better than ever. The hole in the mountains is still there.

Poor Poggibonsi. During those first few weeks we flew missions to rail and highway bridges at Perugia, Arezzo, Orvieto, Cortona, Tivoli and Ferrara. Most of us had never heard of any of these places. We were very young, and few of us had been to college.

When we weren't flying missions, we went swimming or played baseball or basketball. The food was good—better, in fact, than most of us had ever eaten before—and we were getting a lot of money for a bunch of kids twenty-one years old. Like good soldiers everywhere, we did as we were told. Had we been given an orphanage to destroy (we weren't), our only question would have been "How much flak?" In vehicles borrowed from the motor pool we would drive to Cervione for a glass of wine or to Basitia to kill an afternoon or evening. It was, for a while, a pretty good life. We had rest camps at Capri and Ile Rousse. And soon we had Rome.

Within less than a week, friends were returning with fantastic tales of pleasure in a big, exciting city that had girls, cabarets, food, drinking, entertainment, and dancing. When my turn came to go, I found that every delicious story was true. I don't think the Coliseum was there then, because no one ever mentioned it.

For the most part, the missions were short—about three hours—and relatively safe. It was not until June 3, for example, that our squadron lost a plane, on a mission to Ferrara. It was not until August 3, over Avignon, in France, that I finally saw a plane shot down in flames, and it was not until August 15, again over Avignon, that a gunner in my plane was wounded and a copilot went a little berserk at the controls and I came to the startling realization—Good God! They're trying to kill me, too! And after that it wasn't much fun.

Joe knew no fear until his 37th mission. "Until then, it was all play. I was so brainwashed by Hollywood's image of heroism that I was disappointed when nobody shot back at us." But after that traumatic mission, "all I wanted was out."

"I'm afraid."

"That's nothing to be ashamed of," Major Major Major Major counseled him kindly. "We're all afraid."

"I'm not ashamed," Yossarian said. "I'm just afraid." *(p.101)*

THIS PAGE DECLASSIFIED IAW EO12958

Flight schedule for May 24, 1944. In 8F flew pilot Capt. George Wells (Capt. Wren). In 8A flew bombardier 2nd Lt. Joseph Heller and pilot Lt. Joseph Chrenko (Hungry Joe).

. . . most of us in Corsica had never heard of Avignon before the day we were sent there to bomb the bridge spanning the River Rhone. One exception was a lead navigator from New England who had been a history teacher before the war and was overjoyed in combat whenever he found himself in proximity to places that had figured importantly in his studies. As our planes drew abreast of Orange and started to turn south to the target, he announced on the intercom, "On our right is the city of Orange, ancestral home of the kings of Holland and of William III, who ruled England from 1688 to 1702."

"And on our left," came back the disgusted voice of a worried radio gunner from Chicago, "is flak."

Mark Twain may have nailed it when he espoused, "God created war so that Americans would learn geography."

We had known from the beginning that the mission on Aug. 3 was likely to be dangerous, for three planes had been assigned to precede the main formations over the target, spilling out scraps of metallic paper through a back window in order to cloud the radar of the anti-aircraft guns. As a bombardier in one of these planes, I had nothing to do but hide under my flak helmet until the flak stopped coming at us and then look back at the other planes to see what was happening. One of them was on fire, heading downward in a gliding spiral that soon tightened into an uncontrolled spin. I finally saw some billowing parachutes. Three men got out. Three others didn't and were killed. One of those who parachuted was found and hidden by some people in Avignon and was eventually brought safely back through the lines by the French underground.

Southern France had finally been softened by operations like the highly effective Operation Strangle, a campaign that had throttled German communications in central and northern Italy. The Germans were choked into abandoning Rome and retreating northwards toward a new network of fortifications in the Apennines. Meanwhile, the Group's operations in the Battle of the Brenner were almost a carbon-copy campaign. Again the object was to pinch off supplies from the Germans in their near-impregnable Gothic Line. Besides chopping up this passionately defended railroad artery from the Reich to the Po valley, the campaign also immobilized the enemy in Italy where in April, 1945, he was forced to either surrender or die.

MISSION #78: AUG. 14, 1944

Flew with 489th as formation commander on coastal at Cap Camerat, France. Things really look as though the big invasion will come tomorrow.

The day was a Tuesday. It was August 15, 1944. While the day started in a routine manner, George was aware of coming change.

On this day of the invasion of Southern France both George and Joe were in the air. The two barely knew each other yet here they were, lock step. Both had arisen early to perform the cold water shave and bathe, to attend the mandatory briefing, to bump and lurch along on the trucks headed for the waiting planes, to attend to the details, to lumber into the air and then —what kind of day was this to be?

For Joe it turned into his worst nightmare, a life-defining experience. At dawn they were part of those first bombers, of ultimately 2,000, over the invasion coast. While Joe was enduring his 37th terrifying mission, this one again over Avignon, George was also in the air. It was George's 79th mission and his mission book entry reads:

MISSION #79: AUG. 15

Today was "D" Day for Southern France. I led the 1st box from our Group (487) which was the first bombers over the invasion coast. All together there were 2,000 over the beachhead between 6:50 a.m. and 7:20 a.m. not counting the fighters and the troop carriers. There was plenty of action downstairs. Seven aircraft carriers, four battleships etc. The Air Force had a corridor to fly and the Navy had a corridor for ships. The target for my box was gun positions at Cap Drammont, France. The Army was going ashore at 8:00 H. In the afternoon we had a rough mission. Lost three ships and lots of wounded. We flew 132 ships today.

Joe continues on that same day:

On Aug. 15, the day of the invasion of southern France, we flew to Avignon again. This time three planes went down, and no men got out. A gunner in my plane got a big wound in his thigh. I took care of him. I went to visit him in the hospital the next day. He looked fine. They had given him blood, and he was going to be all right. But I was in terrible shape, and I had twenty-three more missions to fly.[2]

Joe could have authored the quote, again from Mark Twain, "I've lived through some terrible things in my life, some of which actually happened."

Both men, almost miraculously, touched down safely on their airstrip. A week later Avignon raised its ugly head and starred, once more, in George's book:

MISSION #83: AUG. 23

Flew as formation commander with 487th on R/R Bridge at Avignon, France. This is considered the roughest target for anyone in the theater. We had an element of chaff and frags and 4 elements of frags which we dropped on the gun positions. It must have had the Jerries in their slit trenches because we didn't get a shot

In June of 1975, *Playboy Magazine* interviewed Heller, and the key Avignon mission was brought up again. This was a milestone moment for young Joe. Consequently, it became a milestone in *Catch-22* as well.

Heller: At first, I was sorry when nobody shot at us. I wanted to see a sky full of flak and dogfights and billowing parachutes. Was like a movie to me until on my 37th mission, we bombed Avignon and a guy in my plane was wounded. I suddenly realized, "Good God! They're trying to kill me, too!" Wasn't much fun after that.

PB: That sounds like the Avignon mission in *Catch-22*, when Snowden, the gunner, is killed.

Heller: It is, and it's described pretty accurately in the book. Our copilot went berserk at the controls and threw us into a dive. Then one of our gunners was hit by flak and the pilot kept yelling into the intercom, "Help him. Help the bombardier." And I was yelling back, "I'm the bombardier. I'm OK." The gunner's leg was blown open and I took care of him. After Avignon, all I wanted to do was go home[3]

…except for the pitiful time of the mess on the mission to Avignon when Dobbs went crazy in mid-air and began weeping pathetically for help.

"Help him, help him," Dobbs sobbed. "Help him, help him."
"Help who? Help who?" called back Yossarian, once he plugged his headset back into the intercom system, after it had been jerked out when Dobbs wrested the controls away from Huple and hurled them all down suddenly into the deafening, paralyzing, horrifying dive which had plastered Yossarian helplessly to the ceiling of the plane by the top of his head and from which Huple had rescued them just in time by seizing the controls back from Dobbs and leveling the ship out almost as suddenly right back in the middle of the buffeting layer of cacophonous flak from which they had escaped successfully only a moment before.

Oh, God! Oh, God, Oh, God, Yossarian had been pleading wordlessly as he dangled from the ceiling of the nose of the ship by the top of his head, unable to move.

"The bombardier, the bombardier," Dobbs answered in a cry when Yossarian spoke. "He doesn't answer, he doesn't answer. Help the bombardier, help the bombardier."

"I'm the bombardier," Yossarian cried back at him. "I'm the bombardier. I'm all right. I'm all right."

"Then help him, help him," Dobbs begged. "Help him, help him."

And Snowden lay dying in back. *(p.50)*

"On another mission to Ferrara," Heller writes, "one I don't think I was on, a radio gunner I didn't know was pierced through the middle by a wallop of flak . . . and he died moaning, I was told, that he was cold. For my episodes of Snowden in the novel, I fused the knowledge of that tragedy with the panicked copilot and the thigh wound to the top turret gunner in my own plane on our second mission to Avignon."[4]

When his gunner was injured, Heller called on his Boy Scout training to bandage the injury, but his fear and the closeness of death infused one of the great scenes in *Catch-22*

He felt goose pimples clacking all over him as he gazed down despondently at the grim secret Snowden had spilled all over the messy floor. It was easy to read the message in his entrails. Man was matter, that was Snowden's secret. Drop him out of a window, and he'll fall. Set fire to him, and he'll burn. Bury him and he'll rot, like other kinds of garbage. The spirit gone, man is garbage. That was Snowden's secret. *(pp.429-430)*

This mission was a milestone for Joe, who later said

I might have seemed a hero and been treated as something of a small hero for a short while, but I didn't feel like one. They were trying to kill me, and I wanted to go home. That they were trying to kill all of us each time we went up was no consolation. They were trying to kill me.

I was frightened on every mission after that one, even the certified milk runs. It could have been about then that I began crossing my fingers each time we took off and saying in silence a little prayer. It was my sneaky ritual." *(Now and Then, p. 181)*

Joe's 37th mission fell on August 15, 1944. Exactly six months minus one day earlier, George flew *his* 37th mission. Was the 37th mission the black cat?   Here is what was recorded in George's small black book following his 37th:

MISSION #37: FEB. 16th

> *Mission—Campoleone, beachhead. Flight leader 2nd flight. Worst mission I've ever been on. Had direct hit right through tail and cut my rudder cables. Was undecided whether to bail out or not. We salvoed our hatch and were all ready to jump, but I finally decided to bring it in. We made it all right and I was never so thankful to hit the ground. The enlisted men wanted to jump but I talked them out of it. We lost one of our planes. Red Reichard and Dean went down with it. Also Lt. Dunaway who came over with us. Seven men in the ship and only 3 chutes were seen to open. The plane was hit in the right engine and blew it all apart. All the fellows think I'm lucky because today was the third time I've brought back a shot-up plane.*

Once Joe had completed the required sixty missions, he had zero interest in volunteering for more. He had done his duty well. Sixty times he had gone up in his confining glass cage and sixty times he had returned intact but in varying degrees of distress. He had survived. He just wanted to go home. Home. He elected to return to that home slowly and, in relative safety, by a water-hugging ship rather than to step inside a plane again. For many years following the war he flatly refused to fly in any type of an aircraft at all. Not surprisingly, and so understandingly, he harbored a very real terror of flying, as did many of the scarred combat veterans.

AFTER THE WAR

Using his GI benefits, Joe went back to school earning collegiate advanced degrees. He wrote, was published, lectured, and taught. In 1982-1983, while working through a divorce from his first wife, he developed Guillian-Barre Syndrome, a life-threatening neurological disease involving partial paralysis, which convinced him he was near death. Joe recovered and remarried. This handsome man with his feral, curly, white mane enjoyed bas cuisine with friends who were used to his crotchety and caustic manner.

At the conclusion of an interview with *Playboy Magazine* in 1975, Joe was asked, "Is there any special way you'd like to be . . . remembered?"

Heller: "Remembered? In order to understand that question, am I to assume you have euphemistically deleted the word death?"

PB: "We were hoping you wouldn't notice."

Heller: "It is impossible to predict or control how you will be remembered after your death. In that way death is like having children. You never know what will come out. In Beckett's *Endgame*, he asks his parents, in effect, "Why did you have me?" and his father replies, "We didn't know it would be you.""

To the pointed question, Joe responded, "I fear death, nursing homes, and vaccinations." Eventually he dealt, saying, "Everyone else seems to get through it alright so it couldn't be too difficult for me."

At age 76, Joseph Heller died of a heart attack.

"Joe Heller is dead but *Catch-22* will live forever. He would have preferred the opposite, but what can you do? Death is the ultimate *Catch-22*." (Peter Carlson, *Washington Post* Staff Writer, 12/14/99)

CHAPTER 3

THE MISSION: THE COMMON DENOMINATOR

This is the true accounting of Mission Glassknob told from the vantage point of a pilot, Walter Wooton, of the 486th squadron who was flying the #4 position aircraft on this mission.

I awoke suddenly, as I always do on mornings when I'm scheduled to fly. I'm not aware of having heard anything. The tent was pitch black, with a slightly lighter triangle at the end where the flap is pulled back. My sleeping bag feels warm and cozy, and I'm wide-awake.

A shadow darkens the entrance, and the Officer of the Day will call very softly, "Woot?"

I answer, "I'm awake," and I hear the crunch of gravel as the O.D. goes on to the next tent to awaken another man scheduled for the morning's mission.

The evening before, George (Wren) and Chief (Havermyer) or Fred (Piltchard) and Joe Ruebel (Danby) would alternate researching and planning the next day's targets.

We always had all of the arrangements for the next mission available each night to work on. At least one pilot, bombardier, and navigator plus gunners, drawn from any of the four squadrons, needed to be assigned to each aircraft. The new arrivals would always be paired with the more seasoned men. Crews were not assigned to the same plane each time. They flew whatever aircraft was available. Usually between 6:30 p.m. and 7:30 p.m. 12th Air Force Headquarters would call and give us the target for the next day. We would receive the location and identification, if it was a bridge, troop area, shipping vessel, road intersection, airfield, etc. We were told the bomb load and weight required and the number of bombs, which in turn determined the number of airplanes needed over the target.

Whoever was on duty that night would immediately call the four squadrons and tell them what the bomb load was to be and the number of aircraft they were to furnish. They would then immediately send their armament people out to load the planes accordingly, so they would be all ready early in the morning.

I unzip the sleeping bag and roll into a sitting position on the side of the cot. Now that it is winter we've taken down the mosquito bars, and getting up is less like fighting your way out of a fishnet.

Somebody had turned on the phonograph up at the Officers Mess. They're playing "The Wiffenpoof Song," which seems pretty appropriate. "Damned from here to eternity, God have mercy on such as we." I feel a shiver run up my spine.

I pull on my ODs, GI shoes, and fleece-lined jacket, then walk up to the mess hut. After the darkness outside, the mess seems uncomfortably bright. They have pancakes, bacon and coffee for us today. I'm not usually hungry before a mission, but today breakfast tastes awfully good.

(George: "The cooks would be informed the night before also as to when they should have breakfast ready—if you could call it breakfast. Our expression for it was 'Slum Gullion.'")

The eighteen officers scheduled to fly this morning's mission are here. The rest of the squadron will eat later, after daybreak. There's very little conversation, I guess it's just too early.

I go to the latrine and wait in line, as usual, there's always a line before a mission.

(George: "At first, we had a slit trench for bathroom use, but later, outhouses were built.")

While walking back down to my tent I notice that the sky on the eastern horizon is slightly less black now. Trigger (Paul Phelps, my tent-mate) is still asleep, so I try to move quietly. I get into my flying suit, put my jacket back on over it, pull on my boots and gather up my notebook, pencil, gloves, and earphones. I kneel and shoot up a short prayer, then go out and over to the Operations tent to check out an escape kit. Escape kits contain gold money, a silk map of the area we are to be in, Benzedrine, and emergency rations. These kits had better be sealed when we turn them back in. I stow the escape kit in my shin pocket along with my cigarettes and lighter, then go outside and climb aboard one of the three 2½ ton trucks parked out there.

The trucks start, and we lurch down the unpaved road about a half-mile and stop outside the briefing hut at Group Headquarters. No trucks from the other squadrons are there. Apparently this is to be the 486th show.

Inside the Quonset, bomb-fin crates are lined up to serve as seats facing a raised platform at the end. At the rear center of the platform is a curtain. It covers a map which will show our route to the target by means of red twine pinned on the map. Briefing starts, as always, with a "time hack." "In thirty seconds ... it will be zero six four four . . . ten, nine, eight, seven, six, five, four, three, two, one, Hack!" All watches are now synchronized to the exact second. Now comes the read-off of times: "Start engines 0728, taxi out 0730, take off 0734, reach IP 0942, time over the target 0946." I scribble these in my notebook as they are read out. I've already headed the page with the flight information taken from our Operations bulletin board last night after the mission was posted: 6Z in the lead, 6W in the #2 on his right wing, 6L in #3 in his left wing. I'm flying 6A in the #4 spot, leading the second Vee, with 6Y in #5 on my right wing and 6C on my left in the #6 position.

No rendezvous time is announced, so there will be no fighter escort. Next come the codes for the day: mission "Glassknob"; wounded aboard "Eagle"; dead aboard "Flower"; tower "Gable." The briefing officer cautions us, as always, to observe strict radio silence until across the Italian coast coming home. As usual the emergency signal is a red flare. We should be back on the ground by noon.

(George: The intelligence officer would conduct the part of the briefing covering the expected enemy action. His favorite expression was "There will be flak of no consequence." We would all reply, "That's easy for you to say.")

Now the curtain hiding the map is drawn back, accompanied by the sound of a sigh. It is a reflex action, I suppose, this group intake of breath when the target is uncovered. The target is deep into the Alps, well beyond Lake Garda, near the Brenner Pass. The Major explains the purpose today. Our flight of six B-25s is to drop twenty-four 1,000-pound semi-armor piercing bombs into a mountainside above a rail cut.

This will cause a landslide, or avalanche, which will bury the rail line under tons of rock and keep the line cut effectively for a long time. This is the rail line from the Brenner Pass. Lately, the Krauts have rebuilt the bridges we've knocked out within days, sometimes overnight. Today's raid will make things a little more difficult for them. The Major is gleeful at the prospect. We're not! It's a long way for only six airplanes with no fighter escort. There are Me-109s and Macci-202s up there. And we know they have a hell of a lot of '88s along that river.

The Mission Commander is to be a Lt. Colonel from Group Headquarters. He will be flying 6Z, with the regular pilot in the right seat. When I checked the Squadron Bulletin board last night I wondered why no copilot was posted for 6Z. That explains it. Number 4 (that's me) is to take the lead if anything happens to number one.

Known flak positions are pointed out on the map. Our flight path is routed around all of them except those over the target area. We'll cross the Italian coast at La Spezia, follow a meandering course to Lake Garda, then turn north and follow the river up to the target. We'll cruise at 9300'. Indicated airspeed 200 MPH. The Mission Commander is to radio a coded mission report when we reach Lake Garda on the way back.

With a "good luck" from the Major, the briefing is over and the pilots, bombardiers, and radiomen leave to preflight the aircraft. At the pilots' briefing the five of us (the Colonel from Group doesn't join us) are given a weather analysis, suggested power setting for climb and cruise, are told to maintain a listening watch on Channel B, reminded to maintain radio silence, conserve fuel, and to be sure to turn on IFF (a radar identification device). We'll each be carrying four 1,000-pound bombs, and a full load of fuel and ammunition. We will need every inch of runway to get off since there's no surface wind this morning. I hope we don't need a few more feet than we've got.

We go out to the trucks and are driven another half-mile to the airstrip, down the taxiway to the equipment Quonset. It is pretty light outside now. We jump over the tailgate and go inside the hut to the bins. I take out my Mae West, check both gas cylinders and valves, then strap it on. Next I check the seals and rip cord on my chute, then shrug it over my Mae West.

My airplane, 6A, *Sahara Sue II*, is parked on the hardstand nearest the equipment hut so I walk over without waiting for the truck that serves a line taxi. I do a walk-around check of the ship with my copilot, Red Allison. Red has already completed the preflight list. The crew is all here, and the six of us sit around and smoke, waiting until it's time to get aboard.

Every few minutes someone gets up and goes over to the weeds beyond the hardstand to relieve himself. I marvel that so much water can be passed by so few. But it's always that way before a mission . . . it goes with the job.

Finally, after checking my watch I say, "Let's turn the props over," and we all take turns putting a shoulder to a propeller blade and pushing it as far as we can until the man behind catches the next blade and keeps the rotation going. We count aloud to six, meaning we have rotated the propeller twice, and the engine three times (gearing is sixteen to nine). This drains any oil that has run down to the bottom cylinders, which might crack a cylinder head when the engine is started. We repeat on the other prop and now it's time to go.

My tension has been mounting steadily since I first got up this morning, but I know it will leave as soon as I get the engines started. It always has, and this is my 46th mission. But right now my stomach feels like I've swallowed a cannonball.

I snap my flak vest over the chute and climb aboard. I hear both hatches slam shut behind me as I settle into my seat, fasten my seat belt, and plug in my throat mike and earphones. Red and I run through the checklist. At 0738 I hit the primer switches . . . throttles cracked, prop control full forward, mixture full rich . . . I shout out the window, "Clear left" and hit the starter switch. The big prop turns over and over, then catches with a roar, throwing a great cloud of blue smoke. I follow the same sequence with the right engine, which starts quickly, and the B-25 trembles as if she is anxious to get moving.

Rinky Doo is #3 today, so I watch for her to come down the taxi strip so that I can fall in behind. Here she comes! I glance at my watch, its 0711. I let off the brakes and taxi our behind 6L. Figler, 6Y's pilot, has slowed down to allow me to turn into the line ahead of him.

We stop near the end of the runway to check the mag and run up the engines. I'm dimly aware that the tension I've felt all morning is gone. The lead airplane, 6Z A.W.O.L is on the runway and rolling. It's exactly 0714. Now 6W starts to roll and 6L moves ahead to the end of the runway and holds. Rinky Doo starts rolling and I taxi out onto the end of the runway. Booster pumps On. 15 degrees of flap, I advance the throttles slowly to 44 inches, release the brakes and we start our run. The control van flashes by on my left . . . we're halfway down the strip. I ease the control column back and get the nose-wheel off. At takeoff power the engines sound as if they are tearing themselves from the nacelles. Good old 6A flies herself off the ground a hundred feet to spare. I jerk my right thumb up and the gear starts up. Red had his hand on the handle waiting for my signal. I reduce the power then Red reduces the RPMs while I start milking up the flaps. We're over the Mediterranean at 75' straining to climb with the weight of the armor plate, bombs, ammunition, fuel and men.

Ahead of me the lead plane has started a shallow climbing turn to the right. The number two and three ships start turning too, leading 6Z so as to slide into position on his wing as he comes around. I bank to the right, keeping my nose aimed just ahead of the number three. I'll be flying formation in reference to number one, but to watch him rather than two and three during the join-up would be inviting a mid-air collision. I get closer to number three since I'm turning inside of him, and get just behind and below before he comes into position on 6Z's left wing. I slide into the number four spot, tucked in close behind and below number one.

Shapes on my left and right, at the edge of my peripheral vision, let me know that 6C and 6Y are in position on my wings. The formation is tight. I can count the rivets in 6Z's belly.

It's physically painful to fly the number four position. Your head is tilted back and you are looking up through the top window behind the windshield. My neck muscles begin to protest after a while. Our squadron has lost more airplanes in number four than in any formation position. This fact doesn't bother me particularly. Although I'm not overly optimistic about my chances of completing this tour, I've never believed that any particular position is worse than another. The Krauts aren't that accurate.

I nod to Red to take over. His left hand closes over my right on the throttles and I release them and the control column, and drop my feet flat on the floor. Allison is good. The airplane doesn't waver during the transition, and he keeps us socked right in there. I shake my head to uncramp my neck, light a cigarette, and make a crew check on the interphone: tail gunner, radioman, top turret, and bombardier, each reports everything okay. I check the engine instruments, then the flight instruments. We're climbing through 8,000 feet at 0729.

I jerk violently at a series of explosions much like a truck engine with no muffler. It's only the top turret testing his guns and the smell of cordite seeps into the cockpit. I hope Red didn't notice my startled jump . . . sounding and appearing calm is the prime rule of this game.

We level off at 9,200 feet. I reach over Red's left hand and pull the prop levers back to 2,100 RPM, making minor adjustments until the engines sounded synchronized. I check the fuel gauges and flip two switches to transfer fuel from the auxiliary tanks, out at the end of the wings, to the large main tanks inboard. I like to transfer the reserved fuel just as soon as we have burned enough out of the main tanks to accept it all. Some of the fellows won't transfer fuel until they've left the target and are on the way home. They believe, correctly, that a full tank is less apt to blow up than a tank full of fumes. But I believe that a hit on the fuel transfer pump or lines is just as likely as one in the reserve tanks, and that extra fuel out there won't get you home if you can't transfer it. Furthermore, the tanks are vented, and if you transfer as early as possible, the fumes should be gone before you get shot at. This question is the subject of one of those running arguments, night after night, back in the tent. Nobody ever convinces anybody on the other side. I'll never understand how the Army overlooked this question. There's a regulation on absolutely everything else.

Allison's neck is bound to be bothering him by now. I grasp the wheel lightly, put my feet back on the rudder pedals, l put my right hand over his left, and take the throttle as he slides his hand away.

Looking fixedly at the lead plane a few feet away I can't see the horizon and am never quite sure of our altitude, whether we are turning, climbing, or straight and level. In formations this tight you don't want to risk letting your eyes stray from the airplane you are flying on.

After Red and I have exchanged the controls another few times, Bray, the bombardier, calls on the interphone, "Five minutes from the IP (that Initial Point where the final turn toward the target is made, and is the beginning of the Bomb Run). At the IP you roll out of the turn on a heading to the target. The bombardier then has to find and recognize the target visually, then get it centered and tracking in the crosshairs of his Norden Bombsight.

(George: "The pilot, who had given control of the aircraft to the bombardier, has, in the cockpit, a gauge on the instrument panel called the Position Direction Indicator, or PDI. This device was a relay signal from the bombardier, which showed the pilot if little changes were needed in direction, to right or left, for the direct line to the target.")

Today we will have 240 seconds to do this. During the final 30 seconds of the run we will be flying straight and level at a constant speed. This half-minute is the most dangerous time. More that half the planes lost during my tour have been hit during the fraction of a minute before the bombs are released, and you even begin to take action of any kind. Just hold it in tight.

I nudge Red and he takes over. I bob and turn my head to ease my neck muscles a last time, reach behind my seat for my steel flak helmet, and put it on. I check the engine and fuel gauges, glance outside at the incredible beauty of these magnificent mountains, and fight off a tremor brought on by the cold—or is it fear? I take the controls from Red and from the corner of my eye see him don his flak helmet and lower his seat to Full Down. His job is to watch the instruments during the bomb run, and he says he can concentrate better if he doesn't see outside too well.

The underside of 6Z's wings flash with reflected sunlight—we're turning on to the IP. Those wings ahead and above are shaded again, and I know that we're headed towards the target. 6Z's bomb bay doors open and I can see the long, fat bombs inside. A puff of jet-black smoke passed by the window, then another, and another. Flak! There's a loud **CHUNG** with another simultaneous sound like a fistful of stones flung against corrugated iron. This means we're hit. Everything feels okay, the engines sound fine. Red would have already told me. I'm aware that my anxiety is completely gone, replaced by an exhilaration beyond anything ever experienced outside of combat. I have a sense of being wholly, completely alive. All my senses are acute. Time seems to slow down.

I can see flak bursting dead ahead, then hear the **CHUNG** of another hit. We are still making small turns, climbs, and dives, and haven't settled down for the last straight and level run. Damn! The Krauts are accurate today! They're putting their '88s in our hip pocket and we are still twisting and turning!

My mind is racing with many thoughts: the sound of the engines, our position in the formation, the intensity and accuracy of the flak. "Yey, though I walk through the valley of the shadow of death, I will fear no evil, for Thou art with me. " Those words always pop into my mind, unasked, on the bomb run.

Now we are flying straight and level. I hold 6A very tight just behind and under 6Z's tail stinger (the tail 50 caliber). Flak is bursting just ahead of 6Z; it's right on our altitude but they're leading us a little too much. Suddenly 6Z's four big bombs break free, and wobbling slightly, fall straight down in front of our nose. The seats thrust upward as our own 4,000-pound bombs release and the airplane responds with a swooshing lift.

Okay, Colonel, let's get the hell out of here, I think, and ease up on the controls anticipating a violent turning dive to get out of the flak. 6Z's bomb door closes—nothing happens! The Colonel maintains our creeping airspeed, then starts a gentle 15-degree turn. (I learn later, from his crew, that he's watching the mountainside, and wants to see personally, the avalanche which is to cover the rail line with tons of rock.) It's the tail-gunner's job to do that and to report to him over the interphone.

There's a very loud **BLAM**! and more black smoke flashes by. Red tugs my sleeve and I steal a glance away from 6Z to him. He's ashen. "Did you see it?" he shouts at me. I shake my head, not understanding, and tilt my head back to hold position. It's not hard to do, we're still doing 200 MPH in a gentle turn. Something draws my attention over to Knighton's ship, 6W, in #2 position on 6Z's right wing. He's above me at about one o'clock. His left fin and rudder, flat olive drab with a big white "6W", slowly turns a glossy black, changing color as I watch. I'm flabbergasted, never having seen anything like this. Now his left propeller slows and comes to a stop, feathered, then I realize that the shiny black color came from the oil pouring out of the left engine and that was blown back on the stabilizer and rudder. 6W skids to the right, smoking and losing altitude. Knighton must be literally standing on his right rudder to keep his good engine from turning into the formation. He slides down and out of sight, leaving a trail of smoke. We've completed 270 degrees of turn and roll out level, plodding along at 215 MPH. At last we're out of range of the 88 batteries and are headed home.

(George: "Then started the long trip back to our home base, often with many planes sustaining damage and injuries. During a mission, if a plane was shot down before our very eyes, or had engine problems, there was nothing we could do except make a record of where it went down. Sometimes damages would prevent a plane from keeping up with the formation and they would have to drop back and make it back alone. We would just keep

trying to make it home safely ourselves. There was no way we could help them. There would be ambulances, fire engines and medical people standing by as we landed ready to assist us in any way. The Red Cross girls would also be there with coffee and doughnuts.")

"Red leans over and slides the earphone off my right ear, gets up close and tells me that "Figler got a direct hit back when I heard the BLAM and has gone down." Most of Red's original crew were now assigned to Figler while he gets combat experience flying copilot with me. He watched from a few yards away when they "bought the farm." I glance at Red again. He looks 20 years older than he did a couple of hours ago.

Two ships out of six . . . twelve good men gone! I thank God it wasn't me, then feel a flood of remorse at the selfish thought.

I tap Red's arm and he takes over. I look out to the left at 6C, catch the copilot's eye, point up to 6Z's right wing, then hold up two fingers. He nods and turns his head to shout something to the pilot. 6C begins to drop down, then crosses underneath me and climbs up into the #2 position. Now we're a diamond formation where a few minutes ago we were two Vees.

I light up a cigarette, then get on the interphone and take a crew check, starting with the tail-gunner. With nobody on our wings he's all alone back there and I know his head is swiveling constantly, looking for fighters. Everybody reports everything okay. Our only damage seems to be lots of holes in 6A's skin. She's pretty well patched up already. A few more won't even be noticed. No one adds any comment to this brief report. The usual banter is missing.

(George: "This might be hard to believe, but the hardest part was when you were not on the mission. Waiting for the ships to return, the five or six hours always seemed so long. When the first flight was sighted, the first thing you did was to start counting the planes and looking for damages. Of course, we would get radio calls indicating injuries aboard and calls from planes that had damage and needed priority on landing. If there were no injuries in damaged airplanes, they would wait until the other planes with injuries got down. Planes whose engines were causing problems would get priority landing because of the possibility of the engine quitting.")

I tune the Command Set to Armed Forces Radio at Caserta so that the fellows can listen to some music on the way home. It doesn't sound very good to me. And while fiddling with the command radio I miss the mission report on VHF. Red hears it, leans over and shouts, "Mike, Fox, Nan, George." The mission was a failure! There was no avalanche!

The rail line is still open.

Every airman had to face his first wartime mission. All were young, all were newly trained, and all were thrust into the unknown. Most were apprehensive; some were palm-sweat terrified. As the following summary of the damage done to the echelon of six aircraft while on a mission over Italy July 15, 1944 shows, they had good reason to be.

The Mission - Unvarnished

```
                    488TH BOMBARDMENT SQUADRON (M)AAF
                    340th BOMBARDMENT GROUP (M)AAF

                                                    APO 650, c/o Postmaster,
                                                    New York, N. Y.
                                                    16  July 1944


        SUBJECT:  Damage to Aircraft.


        TO     :  Commanding Officer, 488th Bombardment Squadron (M).

              1.  The following is a summary of the damage done to six (6) aircraft,
        while on mission over Italy, 15 July 1944.


        Airplane No: 43-4064   (8U)

                    1; Stringer cut on Left side of Horizontal Stabilizer.
                    2. Left Elevator holed.
                    3. Right Nacelle door holed.
                    4. Hole through right and left outboard Flap assemblies.
                    5. Two holes through Tail turret, cutting electric wires
                       and Hydraulic lines.
                    6. Large hole through right Wing, near center.
                    7. Large hole through left Wing, near center.


        Airplane No: 43-27537  (8Z)

                    1. Hole in left outboard Wing panel, cutting main spar. (Wing Change).
                    2. Hole in left outboard Wing, through Oil cooler duct.
                    3. Hole in right outboard Wing, through Oil cooler duct. (Poss. Wing Chge)
                    4. Hole in right outboard Wing, through bulkhead and stringer.
                    5. Large hole in left side of Fuselage at tail, cutting Longeron,
                       Former and Stringer. Went on through left Elevator.
                    6. Left Elevator Trim Tab peppered, also Elevator.
                    7. Many tiny holes in left Rudder.
                    8. Large hole through Fuselage, to rear of rear Escape Hatch, and
                       out left side of Fuselage, cutting Stringer on bottom and two
                       Stringers on side.
                    9. Hole to rear of Bomb Bay door.  Cut Longeron and hit Armor plate.


        Airplane No: 43-27669  (8T)

                    1. Hole through left outboard Wing panel and through Main Spar.
                       (Wing Change).
                    2. Hole through left outboard Flap.
                    3. Hole through left outboard Flap Fairing and through top of Wing.
                    4. Hole through left Engine Accessory Section, left lower Cowling,
                       cutting Cowl Flap Actuating Rod and #12 Cylinder Exhaust Stack.
                    5. Large hole, leading edge of right Wing, just outboard of Nacelle.
                    6. Right Rudder torn.
                    7. Long cut in left side of Fuselage, to rear of rear Escape Hatch.
                    8. Hole cut through Bomb Bay while doors were in open position.

                                    -1-
```

Airplane No: 43-27703   (8W)

    1. Bomb Bay doors blown in byond stops, and when pried out on ground, found to be in good condition.
    2. Three holes in left lower front side of Fuselage. (Skin damage only).
    3. Left Waist Gunner Window cracked, but reparable.
    4. Hole through left outboard Wing.
    5. Hole through left Wing and Auxilliary Tank. Rear Spar cracked. (Wing removal necessary).
    6. Hole through right Wing just forward of Aileron.
    7. Hole in right outboard Flap. (needs replacing).
    8. Hole in right inboard Nacelle Fairing.
    9. Hole in right Engine Cowling (lower left)
  10. Small hole in left Bomb Bay door.
  11. Both Landing Light glasses holed.

Airplane No: 43-27729   (8S)

    1. Hole in left Wing tip.
    2. Two small holes in left Nacelle near Wheel Well.
    3. Holes in left Rudder and Elevator.
    4. Small hole in leading edge of Horizontal Stabilizer.
    5. Hole through rear of right inboard Fairing of right Nacelle.
    6. Hole through right inboard Wing Flap.
    7. Small hole through right inboard Cooler Duct.
    8. Two holes in bottom of Engine Cowling, holing Intake Manifold of #8 Cylinder.
    9. Two holes in bottom of Fuselage just forward of front Escape Hatch.
  10. Right lower Plexi-glass cracked in Bombardier's Compartment, and Bombardier's Sighting Window cracked.

Airplane No: 43-27772   (8E)

    1. Holed Horizontal Stabilizer, left and right side.
    2. Hole in right inboard Wing, and Fuel Cell holed.
    3. Hole through bottom of Fuselage, and out through side of Fuselage.

    2. All damage done to these aircraft, and all holes that were received, were from the bottom.

EDWARD M. YACKO
1st Lt., Air Corps,
Engineering Officer.

With his biting sarcasm and sly humor, Joseph Heller wrapped his novel around these wicked and ever-increasing missions: the number of missions required before reassignment home, the destination of each mission, the degree of danger anticipated with each mission, the interplay of crewmembers during the mission, the capriciousness of luck accompanying the aircraft and, primarily, the prayed-for safe return from each mission. It was the single major player in his book, and it gave birth to the now household phrase, "*Catch-22*," referring to an unsolvable situation.

> There was only one catch and that was **Catch-22**, which specified that a concern for one's own safety in the face of dangers that were real and immediate was the process of a rational mind. Orr was crazy and could be grounded. All he had to do was ask; and as soon as he did, he would no longer be crazy and would have to fly more missions. *(p.46)*

Catch-22's Col. Cathcart, who "oscillated hourly between anguish and exhilaration," (p.186) was the character responsible for continually raising the number of missions to be flown before one could return home. In truth, it fell not solely to Col. Chapman, commander of the 340th Bomb Group, but rather to Gen. Knapp, commander of the 57th Bomb Wing, or higher up the military chain of command, to determine the increase of these missions, which grew from 25 to 50, and then, under rare and special circumstances, to 70. It was found that the original number to be flown went so quickly that new crew replacements were put at greater risk by not having enough seasoned crew to fly with and learn from. The carefully considered increase in mission number insured that every flight with new crew would consist of experienced crew as well, decidedly increasing success statistics.

The description in *Catch-22* of the beginning of a mission varied little from the actual facts.

> He [Appleby] was in the Jeep with Havermeyer riding down the long, straight road to the briefing room, where Major Danby, the fidgeting group operations officer, was waiting to conduct the preliminary briefing with all the lead pilots, bombardiers and navigators. *(p. 47)*

"I think," said George, "that I was both Wren and Appleby. Heller describes Appleby as a lead pilot with his lead bombardier, Havermyer. That's me with my lead bombardier, Chief Myer. We flew more as a team than with others. I was considered the hot pilot in the squadron at that time. I was aggressive—always ready to take a challenge. I've always been patriotic. It was easier for me to be a combat pilot than it was for some other people. Like Chief Myer, (Havermyer/Hungry Joe) I think Heller split me into two people."

> [Yossarian] had Orr's word to take for the flies in Appleby's eyes.
>
> "Oh, they're there, all right," Orr had assured him. About the flies in Appleby's eyes . . . although he probably doesn't even know it. That's why he can't see things as they really are."
>
> "How come he doesn't know it?" inquired Yossarian.
>
> "Because he's got flies in his eyes," Orr explained with exaggerated patience. "How can he see he's got flies in his eyes if he's got flies in his eyes." (p. 46)
>
> Everything Appleby did, he did well. Appleby was a fair-haired boy from Iowa who believed in God, motherhood, and the American Way of Life, without even thinking about any of them and everybody who knew him liked him.
>
> "I hate that son of a bitch," Yossarian growled. *(p. 18)*

(George: "By the way, I don't know what Heller means for 'flies in your eyes'. If I had them, no one ever told me about them.')

> The officers of the other five planes in each flight arrived in trucks for the general briefing that took place thirty minutes later. The three enlisted men in each crew were not briefed at all, but were carried directly out on the airfield to the separate planes in which they were scheduled to fly that day, where they waited around with the ground crew until the officers with whom they had been scheduled to fly swung off the rattling tailgates of the trucks delivering them and it was time to climb aboard and start up. Engines rolled over disgruntled on lollipop-shaped hardstands, resisting first, then idling smoothly awhile, and then the planes lumbered around and nosed forward lamely over the pebbled ground like sightless, stupid, crippled things until they taxied into the line at the foot of the landing strip and took off swiftly, one behind the other, in a zooming, rising roar, banking slowly into formation over mottled treetops, and circling the field at even speed until all the flights of six had been formed and then setting course over cerulean water on the first leg of the journey to the target in northern Italy or France. The planes gained altitude steadily and were above nine thousand feet by the time they crossed into enemy territory. One of the surprising things always was the sense of calm and utter silence, broken only by the test rounds fired from the machine guns, by an occasional toneless, terse remark over the intercom, and, at last, by the sobering pronouncement of the bombardier in each plane that they were at the IP and about to turn toward the target. There was always sunshine, always a tiny sticking in the throat from the rarefied air. (pp. 47–48)

At the IP the pilot and copilot took their hands from the controls and the bombardier took over for the next four to seven miles. He had to ignore his survival DNA to ensure that the plane flew straight and steady, avoiding all evasive action, until they reached the exact point to release the bombs. For those excruciating final minutes the crew had to take what was given them for there was absolutely nothing they could do.

If they escaped unharmed the atmosphere for the return home changed. The swatter had not smacked them lifeless and relief oozed from every pore as clenched jaws relaxed and dry mouths moved tentatively once again. Talk resumed. If German guns had taken down any of their buddies in their flight, the quiet interior reflected numbing anguish, each man trying to deal in his own way. At the edge of the runway, HQ staff and ground crews would be anxiously waiting for their return. Eyes strained as incoming planes were counted and as they searched for the dreaded gaps in their ranks.

CONFIDENTIAL

BOMB DAMAGE SURVEY

57 BOMB WING (M)

CONFIDENTIAL

Bombs Away!

View from following plane showing bombs being released.

"You're inches away from death every time you go on a mission. How much older can you be at your age." *(p. 38)*

The mission overshadowed everything. Like Heller, every man remembers his first mission or his most intense mission or his most terrifying mission. For Jack Marsh his vivid memories include the all-important, and ever-increasing, mission count. Marsh was trained as both a bombardier and a navigator and had been assigned to the 340th bomb group on Sept. 1942, 20 months before Joseph Heller's arrival. His circumstances, however, could have been a blueprint for *Catch-22*.

When we first arrived for combat duty, we were advised that after 25 combat missions we would be eligible to return to the States. I was the first person in the 340th to complete 25 combat missions, but was regrettably informed that the minimum number had been increased to 35 missions, NOT 25. Well, as it turned out, I was the first airman in the group to complete 35 missions, but again, we were advised that we would need to fly 50 missions before being eligible to return to the States.

Again, I was the first of the group to complete the 50 combat mission required.   During the six-week delay after my orders were being cut, the US invaded Italy. At the request of my commander I voluntarily joined him in the lead ship for my 51st mission out of Catania, Italy, which was approximately 100 hours, but as we landed back at Catania, we were told not to leave our planes. They were being reloaded with bombs and ammunition, which meant we were expected to fly a second mission back to the first target area. SO, I ended up flying mission number 52.[1]

This continual increase was, of course, *Catch-22*'s lifeblood. Even into his sixties, Joe remembered George very well: "George Wells just flew missions endlessly and without fear, and I put that into *Catch-22* . . . it struck me as kind of heroic and . . . unreal." (The Greatest Generation, Tom Brokow)

Each mission brought to the table its own unique thumbprint. There are as many stories as there are flights. Each change of circumstance required a change of plan by the crews, not by choice but by necessity.

When a cook runs out of butter, a change or a substitution can be made. When a teacher falls ill, a substitute will step in. In like manner, when two crippled pilots ran out of individual functioning parts they substituted with what ingredients remained.

Meet Gat and Otis:

They were just over the target south of Bologna and the flak was flying thick enough to drop their gear and spin their wheels on. The plane on their left was hit; the plane under them went down. The air was bursting wide open on all sides. Suddenly Gat (Lt. Charles D. "Gat" Ross, first pilot) got a terrific impact in the right leg and a dizzying jolt on the right side of his face. One piece of flak had torn a hole in the calf of his right leg and another had ripped through his lip and cheek, imbedding in the flesh just under his cheekbone. Blood spurted all over the plane, blinding him.

At the same time, Little (Lt. Otis Little, copilot) was hit in the right hand but Gat motioned for him to take the controls, and he flew with one hand holding the formation.

Opening the window, Gat began to get his senses back when the rush of cold air hit his teeth. He thought all his teeth had been knocked out by the flak, or else his jaw crushed for he could feel the jagged edges of the sharp metal. Reaching up, he pulled the piece of flak out of his face, felt relieved, and turned around just as another piece of jagged metal whammed through the cockpit, slashing Little across his left ankle.

Shortly afterwards, one motor went bad. They feathered the prop and limped along, a battered plane on one engine, carrying a pilot with one useless leg, a slashed face, and half blinded by his own blood; and a copilot with one useless hand and one useless leg. There was nothing to do but combine their assets.

So that's the way they brought her home. Gat's hands were all right, so he handled the wheel, while Otis, whose eyes were not bothered by any blood, helped guide him. The pilot used his uninjured left foot on the left rudder pedal, and the copilot duplicated on the right pedal with his uninjured foot.

Nearing the field, the blond, stocky pilot called the tower and told them to "Get the meat wagon ready," that he was coming in on one engine and two wounded guys. He didn't say which two.

Neither one knows how they did it, but the synchronization special was a "greaser," this outfit's term for the smoothest landing possible. Then Ross and Little (who eventually did not survive the war), were hauled off to the hospital.[2]

Pilot and co-pilot in the cockpit

Missions are to be endured and then, hopefully, rehashed over the years at get-togethers. On the evening before one of these future reunions, lights burned way late in one of the member's rooms in the club. Two 340th veterans, who happened to be born on the same day, May 16th, were recapturing an experience. Pilot Harry George and Bombardier Ed

Dombrowski were talking in cadence to others about an exploit.

"We were both on the same mission," Harry said. "We got the hell shot out of us."

The missions on which only one of them flew were milk runs.

"We hoped," Ed said, "we'd never fly together on the same plane."

But on June 22, 1944, they did—for the first and last time.

"I'd never fly with him again," Harry said.

At 7:35 that evening in 1944, their plane was shot down between the Italian cities of Florence and Bologna over the Apennine Mountains. Burned from the flames, they parachuted from the aircraft.

"Ed came down on one side of the mountain," Harry said, "and I came down on the other."

Ed was captured by the Germans, escaped and fought with partisans during his 59 days in Italy. "Harry had love and broads on his side of the mountain," Ed ribbed his friend. Harry smiled. (He and his family in 1969 visited the mountain town of Barerino de Muegello whose residents had hidden him from the Germans during his 78-day stay.)

But Ed had been declared dead. When he arrived home in Erie, Pa, Ed recalled, "The day I got back they were having a High Requiem Mass for me."

On the wing roster, Ed Dombrowski was listed as dead.

Then, in 1976, Ed showed up at the Wing's 7th reunion. At the registration desk, Harry's wife signed him in and burst into tears.

"It was quite a reunion," Harry said, "after believing he was dead for 32 years."

There were occasionally stand-down days between missions. One such dull day occurred because of foul weather and "Chief" (Havermyer) convinced George (Wren) and Fred (Pilchard), and Cal Moody (Moodus) into having a flare gun battle with the British Liaison Officer and his assistants. "Have you ever seen a flare hit a tent? To say the least—we were lucky we didn't start a fire we couldn't control or hurt someone. It was also fortunate Chapman was gone for the day."

On another such day, after having arranged for the British Liaison Officer to go on his first combat mission (no defenses expected, truly a "milk run"), "Chief" and George took down the Britisher's tent, while he was on that mission, and marked his gear for "Next of Kin." "Thank heavens he returned safely. We did help him put his things back together again."

Another mission, each one being unique, connected the tiny island of

Pantelleria with the life of Col. Willis F. Chapman, commander of the 340th Bomb Group. This 32-square-foot dot on the map is an Italian island in the Strait of Sicily in the Mediterranean Sea. It lies 62 miles southwest of Sicily and 43 miles east off the Tunisian coast. Its capture was deemed paramount to the Allied success in invading Sicily since it would permit planes to be based within range of that larger island. This tiny but vital strip of land had been the focus of continual, intense, unrelenting, day-and-night pounding by fighters and bombers before its unconditional surrender was tendered, the surrender that coupled it with Bill Chapman.

In Bill's words:

The Army had planned to take the surrender of Pantelleria at noon on a Saturday. Pantellerians had been given their instructions, "No firing! " A big cross was to be laid out on the airstrip if everything was satisfactory. We tried to find how to land on that strip. General Auby C. Strickland, Desert Air Task Force, said he wouldn't touch it with a ten-foot pole. Nobody could get in and out of that pock-ruined field.

Strickland said to me, "If you get in and out of there and are still alive, you come on over and I'll give you a free dinner. "

"What kind of dinner? " I asked.

"Lobster. "

"OK, I'll be over. "

I studied the area's photos and found two places where I could land but they weren't straight. You had to hit and zig and zag back a little. I could make it. They gave me the papers and, if I was out at 5 o'clock on Friday and all was clear, the Pantellerians would surrender. But whom would they be surrendering to? To the Air Force! (The AF planned to capture it as a test case for air power.) It would be the first time in history that any outfit had ever surrendered to the Air Force. It was very important to the Air people; they wanted it so bad they could taste it.

I came in a few minutes ahead of 5 o'clock. I figured if I hit a pothole there would be someone there to fish me out. If I made it I could take the surrender. I made it all right and got to the hanger —but not a soul was on that field. Not a one. Nothing there. Things aren't right; maybe I'll get shot, I thought.

They had underground hangers that I explored. These labor-intensive hangers had been hewn from solid rock, thereby ensuring their safety from bombardment. Unfortunately I didn't bring any blankets and it was a time of year that got damned cold at night. I couldn't shut the hanger doors. I remember sitting there and freezing all night long. Thought in the morning I could have C-rations I'd stuffed in the duffle bag. Lamb stew. In the morning I had it for breakfast. Opened it, took one bite and went AGH! And threw it away. Don't know how the rest of our people ate it. It was about 5-6 a.m.; the Army would come at noon. Not a damn soul to give the surrender.

I checked out the airstrip to see how I could take off. There were two aircraft on the field—one, a British Hurricane sporting the British Union Jack, was standing on its nose in a pothole. The other was also British, a Beaufighter. (The "Beau" was a long-range heavy fighter well regarded by its crews for it ruggedness and reliability. By the end of the war 70 pilots with RAF units had become aces while flying this aircraft). I checked out the airstrip to see how I could take off. I plotted it out, cranked her up, dodged all the right holes at the right time, and I flew back to Tunisia.

Bill eventually collected his promised dinner. As he lifted the shiny dome from his plate, there sat, in all its glory, a can of Spam! It, however, shortly was followed by a fine orange-pink langoustine, the smaller edible lobster of the European seas.[3]

Ahhh . . . a mission with a happy ending!

BILL:
SOLID
COLONEL CATHCART
SPLINTERED

Bill Chapman had a choice. Headquarters MAAF (Mediterranean Allied Air Forces) offered him command of either the highly successful 321st Bomb Group or the "Unlucky 340th" Bomb Group.

It was March 1944, two months before Joseph Heller's arrival on Corsica. The 340th Group had been suffering. Its morale was at a low point due, in part, to the fact that it had lost two of its past three commanders, one shot down and killed on a mission and the second shot down and taken prisoner for the duration of the war. Its lack of bombing accuracy was notorious. Their B-25s were desert war-wearies complete with desert camouflage. They were C & D models of the B-25 and were outfitted with outdated, manually released British Mark IX bomb sights that were preset and allowed no fine-tuning. Only one aircraft in the outfit sported the highly desired and fanatically guarded American Norden Bomb Sight, the king of accuracy.

Bill chose the 340th.

On 15 March 1944 I assumed command of the 340th Bomb Group, and on 16 March lava started flowing down from the north side of Mt. Vesuvius. It was predicted that a church steeple in a small town, in the path of the lava flow, would fall down at 1606 hrs. on 21 March.

Major Joe Ruebel, assistant operation officer, and I drove on the 21st from our Pompeii base, on the south foot of Vesuvius, around to the north slope just in time to see the church steeple collapse exactly as predicted. We watched, for a short time, the town being slowly engulfed by a molten wall of slow-moving lava about 100 feet thick. The lava wall was black except for when the front sheared off as it moved, revealing a red molten interior, which rapidly turned black as it was exposed to the air. The power required to move such a mass is unbelievable.

As Joe and I returned around the east side of Vesuvius, toward our airdrome, Vesuvius blew her top (1725 hrs.). Black crud and corruption rose to what we estimated as 20,000 feet before it spread, severely restricting visibility in all directions, particularly toward our base. The wind coming from the northwest blew the airborne debris to the southeast, over the 340th and on for 200 miles, toward Bari on the Adriatic side of Italy. It created its own fog and darkness under this stream of lava dust and clinkers up to basketball size. Within minutes any thought of getting aircraft airborne was impossible. We couldn't even find them. It was also impossible to assess accurately the damage, which was accumulating rapidly.

In addition to keeping the staff of the 57th Bomb Wing advised of our deteriorating situation and needs, I called my old office at HQ MAAF in Caserta about 2300 hours, and by daylight we had US Army trucks streaming in from many places. Some even came from the front lines at Monte Cassino.

We managed to salvage just about everything except our now-worthless shredded canvas and our aircraft. They were severely damaged and mired in 18-20 inches of ash. We moved our operations to the Guado airfield at Paestum, Italy, south of Salerno. With the B-25s borrowed from the 321st, the 340th flew a 24-ship mission just four days after Vesuvius erupted.

In hindsight, this violent and scary event was a blessing in disguise. No one was hurt. We got rid of all of the desert camouflaged B-25Cs and Ds fitted with British bomb sights, and received rapidly from the Replacement Depot in Tunisia new B-25Js with the super-secret Norden sights, which were immensely more accurate. This was a big turning point in the fortunes of the 340th Bomb Group.[1]

The famed and highly secret American Norden bombsight made pinpoint bombings possible. 1944.

General Robert Knapp recalled how the Norden was so highly protected that in class, instead of a manual on its operation, the students took notes and sketched from the blackboard. At the end of the class, the instructor retrieved the notes. The next day the scenario repeated. By the end of the course the students had memorized its operation. Bill also offered a memory, that of the B-25. It was to describe that aircraft as "the noisiest, the damnedest, the coldest, the hottest. But it was the sturdiest. It was the best."

Bill Chapman was the right man for his job. His birth, 21 years after his brother, Carl, died in babyhood, came as first a shock, then a blessing. Arriving so unexpectedly, and in truth, somewhat embarrassingly, to parents in their forties, Bill was doted on. His father, William Henry Chapman, was a Michigan railroad engineer who, although he dealt with a powerful coal-fed, black-soot-producing locomotive on his runs, always returned to their charming Victorian home on the corner of tree-lined Orange Street, meticulously clean after showering and changing into fresh clothing before leaving work. His white nails belied his occupation. His mother, Minnie Lee, of Robert E. Lee ancestry, often taught piano. Their genes combined to give this handsome, dark-haired son a creative mind with an enviable ability to focus.

Bill set his goals and then accomplished them. From a young age he shouldered five years of a paper route in his Jackson hometown. He became an Eagle Scout, a summer camp counselor, and he attended Jackson Junior College. He worked in the library after school, where he honed his lifelong passion for reading (at his passing, his fifty-two current magazine subscriptions affirmed this fact) and, in the stacks, discovered the West Point Military Academy. Bill's many military ancestors stretched back to America's Revolution and Civil War so, even though he had no personal knowledge of anyone who had followed this path, his pairing with West Point was, perhaps, destined.

Entering West Point, with its unbending traditions and expectations, he found that the instructors provided ample opportunities to round out their cadets. He set goals. One became the men's chorus. Bill saw how the West Point chorus took trips to perform. He wanted trips.

When the try-outs for the chorus opened, he joined the long line of waiting cadets, each dressed with painstaking fastidiousness in his pin-neat, grey-black-and-gold wool uniform, to audition. The choir director, Mr. Myers, sat at a piano, and each auditioning cadet would sing "Glory to God!" Four brief syllables and the pronouncement was either Yes or No. OK or Next.

As Bill recounted, "I had been in the church choir several years. I hit my best note and I thought this was a breeze. There would be no problem. There was a problem. I got thrown out. The guy behind me had the biggest voice of all. He was a bass and I was a second tenor kind of thing. He let out a beller—didn't sound like any tune at all but he really made a lot of noise and Myers said, 'OK.' I thought this was a hell of a note, so I went around to the back end of the line again and this time I let him have it, and he said 'OK'."

That second time around, Bill's resounding, full-throated, strong voice boomed it out—volume they wanted, volume he would give them, and a spot in the chorus they gave to him. And those two weekend trips to New York.

The choir also permitted Bill a bit of independence. Instead of joining the required formation that marched cadets to their seats in church and, again, marched them out after the final Amen, he, now being a chorus member, had privileges. "I could walk up like a human being. And I could walk out."

Then there was the football team. "Years earlier I had injured my knee and the doctor said I would not be able to do running anymore and he recommended swimming. So I swam for a long time—at camps, intermediate school, high school, and junior college. Then I discovered I was doing better at running. West Point did not know I had that injury or they wouldn't have admitted me in the first place. I got word from upperclassmen that if you got on the football team, you got to eat at the training table. Sold me on that. You did not have to sit at attention and could eat all you wanted. I could eat five, six, seven steaks at a time. If I worked hard at it during practice, I could knock off thirteen, fourteen pounds. On the field all that water sweated out and we would come in ravenous." He maintained his trim build and played a mean football tackle for four years to have access to that coveted training table.

This young man, with his supposedly wounded and limiting knee, also took in stride four years of swimming and track as well as the less physically challenging Chess Club and Cadet Orchestra, while becoming a rifle and pistol sharpshooter. A striking, engraved presentation saber for ranking first in engineering drawing attested to his attention to academics. It was a fast course, he remembers. "You didn't stop to sharpen your pencil."

As his 1935 West Point *Howitzer* yearbook inscribed, "In Bill you see a man who accomplishes what he sets out to perform. Give him enough time to make out his 'poopsheet,' and he asks odds of no man."

The 340th Bomb Group was fertile ground for such a commander, and its turning point accelerated as he began initiating necessary changes to the performance of missions as well as to the men's confidence in their abilities.

Bill immediately recognized the need in this somewhat shaky "Unlucky 340th" for an extreme morale makeover. He would start at "home." He had a large, bold, and emphatic wooden sign built on site and installed near the base entrance, which read in spine-stiffening and prophetic letters:

Operations
340th Bombardment Group
(Medium)
The Best Damned Bomb Group There Is
Product of the USA

Reminiscing about this structure, George recalls, "He knew when a visitor came to see us, that we would strive to live up to those words, and he knew when we were out in front of that sign, on a mission or whatever, we would do our darndest to make those words true—and we did!"

Willis F. "Bill" Chapman, right, signing in after a mission.

Bill quickly swung his attention to improving the group's bombing accuracy. This was where success was imperative and efforts could not be lukewarm. Germany had to be put on a defensive rather than offensive stance. The 340th's piercing focus had to be on how to break the vital lines of supply and communications with the German army and its allies and then to continually keep them broken. Germany could be crippled, but not without solid work, increasing talent, and a passionate drive to succeed, which would require of each man maximum effort. The focus was to keep Germany off balance and struggling to repair, to catch up.

The 340th tightened its belt as Bill honed in on improvements. He started by assigning extra schooling to the lead crews. They were to practice dropping Blue Bombs on an island off Corsica. Blue Bombs were sand-packed practice bombs weighing 100 pounds each. Lead crews were now required to drop twelve Blue Bombs after every mission. Accuracy improved. Two years previous, the 340th had provisions for a skeet range in their aerodrome; however, those provisions had never been implemented. Fodder for Bill. Shortly after his arrival, he set it in operation. As with the Blue Bombs, the purpose of the skeet range, now a weaponry device, was to increase accuracy. Up to this point only fighter pilots had access to practice training, while the gunners had virtually nothing.

The now-functioning range gave these gunners a vehicle to increase their familiarity with some of their weapons in a non-combatant situation. This range served multiple purposes. The men found it interesting and pleasurable as their skills increased. They enjoyed the mixing and competing together—pilots and gunners, enlisted men and officers. During this period the 340th skeet range served as an enjoyable yet competitive supplemental means to a deadly end. And accuracy improved.

Now, this was just too good for Joe Heller to pass up. In *Catch-22*'s version, Joe indulges General Dreedle, who defined the words of Mark Twain, "To a man with a hammer, everything looks like a nail," and cheers on *Catch-22*'s Lieutenant Dunbar, "A true prince," says Yossarian. "One of the finest least dedicated men in the whole world."

> General Dreedle had thrown open Col. Cathcart's private skeet-shooting range to every officer and enlisted man in the group on combat duty. General Dreedle wanted his men to spend as much time out on the skeet shooting range as the facilities and their flight schedule would allow. Shooting skeet eight hours a month was excellent training for them. It trained them to shoot skeet.
>
> Dunbar loved shooting skeet because he hated every minute of it and the time passed so slowly. He had figured out that a single hour on the skeet shooting range with people like Havermeyer and Appleby could be worth as much as eleven-times-seventeen years.
>
> For Dunbar his perilous life seemed to stretch if it were filled with periods of boredom and discomfort. *(pp. 37-38)*

With its eye still on the accuracy prize, the 340th's increasingly creative mind fashioned the homemade "Squirt Gun Turret," another war toy. The squirt gun turret worked like this: A post, anchored in the ground, carried a 5- to 8-foot arm with a model airplane attached to its furthermost end. A grease gun with a firing range of 15 to 20 feet was filled with water and mounted on a pivot. To enable the gunner to determine direct hits, the water was tinted green. To increase the degree of difficulty, the airplane's speed was continuously altered and a cam was added to the arm to infinitely vary the flight pattern. The object of this exercise was to emphasize the importance of leads: how much advance lead to give an aircraft flying at certain heights and speeds to insure a direct hit. And accuracy improved.

The squirt gun turret soon became a competitive diversion, generating an intersquadron competition in the midst of the war. It was but another example of the continually stressed importance for the men to keep their minds turned to improving, improvising, and inventing survival techniques. Always they were encouraged to do the best they could with what was available, to keep minds open to new ideas; to try it—it might just work.

Full view of two men on
Squirt Gun Turret

DAY - 30 TH
STAND DOWN

Sighting the target.
The unique "Squirt Gun Turret Trainer" was invented by the 340th Bomb Group.

Bill now stepped into more difficult and much more dangerous waters, waters that could suck under the very lives of him and his men. He originated a military strategy that was referred to as the "straight and level bomb run."

Earlier on, pilots, after a short turn toward the target, had relied on their maximum evasive and protective maneuvering skills—the banking, climbing, dropping, and rolling—to confuse enemy anti-aircraft units, before leveling out for the 5–10 seconds needed to drop their load. This, of course, was safer —but not effective. Now change was in the air.

As a mission took flight it was composed of from 6 to 78 aircraft destined for a specific bombing site. These aircraft were clumped tightly in boxes of six ships. Following the lead aircraft, the other planes would dip, by the thickness of one plane, below the one before it. This enabled those pilots to better see ahead and made the target considerably more difficult for the enemy.

Here is where Bill's innovation came into play. In this box of six, one of the ships was designated the lead or command ship. Now, for up to five minutes before the target was reached, the lead bombardier's duty, as he took control of the plane from the pilot, was to fly level and steady. These were men like Joe and Red and Chief. For these perilous minutes he had to hold this ship on course, remaining on that site continuously and precisely in this tight pattern. This extra undisturbed period gave him valuable time to adjust and fine-tune. He was instructed to fly this level and steady pattern once the sights were fixed and take out the target no matter what the danger. And accuracy improved.

The aircraft were flying on those three different levels to increase the difficulty of being hit, but it took close teamwork and steady, controlled nerves because, most often, the dreaded flak—those jagged chunks of iron fired from anti-aircraft German ground crews defending the target—was in the air and eager to indiscriminately maim and kill en masse.

The planes were tucked so closely together that the faces of the crews in adjacent planes were clearly visible to each other and, even in the realm of huge danger, these air crews often felt the thrill of flying in a tight and imposing formation to and from their target.

George's 96th mission became intimate.

96th MISSION —DEC. 30TH, 1944
Flew as Flight Leader with 488th Sqd. on target at Calliano. R/R Bridge #1. Lots of ack-ack and when I broke off the target I looked back and had to laugh at Jinks* who was heading the 2nd box. Capt. Myers (Chief) was my bombardier and he hit the 120 ft. bridge dead center from 12,400 ft.

The Group became known as "The Bridge Busters" because of their accuracy. For his heroic piloting on this mission, George was awarded the Silver Star.

For gallantry in action. On 30 December 1944, Major Wells led an eighteen-plane formation in an attack upon a heavily defended railroad bridge near Calliano, Italy. Upon the approach to the target, intense anti-aircraft fire enveloped the formation, heavily damaging his airplane and ten other B-25s. Determinedly maintaining his crippled plane in lead position in the face of this accurate barrage, Major Wells enabled two flights in his formation to register many direct hits, heavily damaging this vital bridge. To insure maximum effectiveness of his mission, Major Wells then guided the formation

---

*   *Pilot John "Jinks" Turnbull, and George had been friends since their National Guard training days. Jink's direction took him to the amphibious planes but he found he did not like it and George was able to get him transferred. After the above incident, Jinks wrote to George's parents describing how George had laughed at his getting shot at. They admonished their son, lightly scolding him in their next v-mail, that he really should not do that. George ribbed Jinks, "This is what you wanted."*

on a perfect run over the alternate target, thereby enabling his third flight to release its bombs with devastating effect upon the objective. An Assistant Group Operations Officer, formation leader and staff observer, Major Wells has participated in 100 combat missions during his present tour of duty. His exceptional ability as pilot and his outstanding leadership have made a marked contribution to the efficiency of his unit. His gallantry in action and steadfast devotion to duty reflect great credit upon himself and the Armed Forces of the United States.

Mission #99 notes Jinks presence again:

99th MISSION, FEB. 21
Command Pilot with 488th on the R/R Bridge at Bressenone on the very top of the Bremer Lines. Jinks was flying as Flight leader in the same ship. We had bad weather and finally had to turn around after a hard time through the weather. Got shot at from Vicenza.

Fortunately Jinks survived the war, as well.

Heller wryly exaggerates in *Catch-22* the progression of the bomb run, first before the straight and level requirement was introduced and then, later, after it was implemented.

Catch-22's evasive bomb runs—
<u>*before*</u>
the straight-and-level bomb runs were implemented

Havermeyer was a lead bombardier who never missed. Yossarian was a lead bombardier who had been demoted because he no longer gave a damn whether he missed or not. He had decided to live forever or die in the attempt, and his only mission each time he went up was to come down alive.

The men loved flying behind Yossarian, who used to come barreling in over the target from all directions and every height, climbing and diving and twisting and turning so steeply and sharply that it was all the pilots of the other five planes could do to stay in formation with him, leveling out only for the two or three seconds it took for the bombs to drop and then zooming off again with an aching howl of engines, and wrenching his flight through the air so violently as he wove his way through the filthy barrages of flak that the six planes were soon flung out all over the sky like prayers, each one a pushover for the German fighters, which was just fine with Yossarian, for there were no German fighters any more and he did not want any exploding planes near his when they exploded. Only when all the Sturm und Drang had been left far behind would he tip his flak helmet back wearily on his sweating head and stop barking directions to McWatt at the controls who had nothing better to wonder about at a time like that than where the bombs had fallen.

"Bomb bay clear," Sergeant Knight in the back would announce.

"Did we hit the bridge?" McWatt would ask.

"I couldn't see, sir, I kept getting bounced around back here pretty hard and I couldn't see. Everything's covered with smoke now and I can't see."

"Hey, Aarfy, did the bombs hit the target?"

"What target?' Captain Aaradvaark, Yossarian's plump, pipe-smoking navigator would say from the confusion of maps he had created at Yossarian's side in the nose of the ship. "I don't think we're at the target yet. Are we?"

"Yossarian, did the bombs hit the target?"

"What bombs?" answered Yossarian, whose only concern had been the flak.

"Oh, well," McWatt would sing, "what the hell."

Yossarian did not give a damn whether he hit the target or not, just as long as Havermeyer or one of the other lead bombardiers did and they never had to go back. *(pp.29-30)*

Catch-22's evasive bomb runs—

<u>*after*</u>

the straight-and-level bomb runs were implemented

He [Yossarian] came in on the target like a Havermeyer, confidently taking no evasive action at all, and suddenly they were shooting the living shit out of him!

Heavy flak was everywhere! He had been lulled, lured and trapped, and there was nothing he could do but sit there like an idiot and watch the ugly black puffs smashing up to kill him. There was nothing he could do until his bombs dropped but look back into the bombsight, where the fine cross-hairs in the lens were glued magnetically over the target exactly where he had placed them, intersecting perfectly deep inside the yard of his block of camouflaged warehouses before the base of the first building. He was trembling steadily as the plane crept ahead. He could hear the hollow boom-boom-boom-boom of the flak pounding all around him in overlapping measures of four, the sharp, piercing crack! of a single shell exploding suddenly very close by. His head was bursting with a thousand dissonant impulses as he prayed for the bombs to drop. He wanted to sob. The engines droned on monotonously like a fat, lazy fly. At last the indices on the bombsight crossed, tripping away the eight 500-pounders one after the other. The plane lurched upward buoyantly with the lightened load. Yossarian bent away from the bombsight crookedly to watch the indicator on his left. When the pointer touched zero, he closed the bomb bay doors and, over the intercom, at the very top of his voice, shrieked:

"Turn right hard!
…just as eight bursts of flak broke open successively at eye level off to the right, then eight more, and then eight more, the last group pulled over toward the left so that they were almost directly in front.
…turn left hard!" he hollered to McWatt, but the flak turned left hard with them, catching up fast, and Yossarian hollered, "I said hard, hard, hard, hard, you bastard, hard!"
And McWatt bent the plane around even harder still, and suddenly, miraculously, they were out of range. The flak ended. The guns stopped booming at them. And they were alive." *(pp.144-149)*

Chunk of heavy iron flak.

With the inevitable losses occurring, Chapman employed the use of chaff[2]. These were strips of metal foil released in the atmosphere from aircraft to obstruct and confuse radar detection. The highly inventive mind of Benjamin Kanowsky, pilot and mess officer, upon whom *Catch-22*'s enterprising Milo Minderbinder was lightly based, was involved in its creation for the 340th. Since the flying of missions was critical and mandatory, everything was done to try to safely shield the B-25s' crewmen. Sometimes there were escort fighters. Chaff, and then phosphorus bombs, were deployed. All were designed to aid in a successful and safe mission.

As the Germans retreated to northern Italy, the Brenner Pass became their most important resupply route. The Battle of the Brenner, begun in November of 1944, continued to be the prime target area for these B-25s of the 57th Bomb Wing. The Brenner also became one of Germany's most heavily fortified areas, with anti-aircraft guns, principally the dreaded "88." As the Germans were being defeated in battles of northern Italy, they placed more and more anti-aircraft guns into position to protect their marshalling yards, bridges, transformer stations, and rail lines. Towards the end of the war they were estimated to have over 1,200 heavy anti-aircraft guns in stationary positions along the Brenner line, plus the capacity to bring in many more that were designed to be mobile and placed upon railroad cars to move from one target area to another. Any crewmember of the 57th Bomb Wing who flew missions against enemy targets in the Brenner could attest to the severity of the enemy anti-aircraft fire encountered.

Chapman pressed on. To protect his men he employed a new anti-flak technique that used highly effective phosphorus bombs against the enemy gun positions. This risky decision was made by the command with full consideration, and despite the Germans having declared that the use of phosphorus against their gun positions was a war crime, and that any members of anti-flak crews who were shot down would be summarily executed. The shoulder-holstered .45 carried by the American fliers took on a new importance.

(From 2nd Lt. Art Curry, a pilot in the 445th: "I remember the news with two feelings: one—while we did carry '45s in shoulder harnesses, that's true, it is also true that it took me three attempts to get my Marksman badge; the second was that the phosphorous bombs must be pretty effective and potent to make the Nazis squeal. Oh, and one other feeling: I sure didn't want to be shot down and if we were, I sure didn't intend to just let them shoot me, bad shot or not." Brave thoughts from a still-twenty-year-old.)

White phosphorus bombs on enemy AA positions at one of the Po River crossings, 1945.

The new plans called for a flight of three planes to precede the main formation and drop phosphorus bombs over the anti-aircraft guns. This solved many problems. Results proved that more lives were spared than lost because of these increased measures; the stunning increase in accuracy reduced the double jeopardy of crews having to return, once again, to a target missed.

The mission successes flourished dramatically. Strikes became pinpoint and consistent.   The 340th was becoming a well-oiled machine of dependability and skill. Bill strove to instill justifiable pride in his men. They were laboring hard; their skills were improving—no, soaring—and they were bombing the hell out of the German defenses.

Two innovations must be mentioned here:

The RADIO RELEASE technique provided for a radio signal from the lead aircraft to release bombs from all aircraft simultaneously. This provided for a much more compact pattern of bombs on impact with greater destructive capability. The Radio Release equipment was designed and developed by two airmen in the 340th under the supervision of the Group Signal Officer. They cleverly utilized some unused circuitry already in the B-25. It worked properly the first bomb run and ever after.

The introduction of SHORAN (Short Range Aid to Navigation). This involved equipment in the lead aircraft which would interpret radio signals from two ground stations which allowed precise distance measurement to the bomb release point. It was possible to do accurate bombing when the target was obscured from the aircraft by a layer of clouds below the aircraft.

-Bill Chapman

With these dramatic changes, Bill sent his Deputy Group CO "Mac" Bailey to Rome to invite well-known Gill Robe Wilson, The *New York Herald Tribune*'s top correspondent, to visit the 340th. Most articles being written glorified only the glamorous fighter outfits. The resulting periodic publicity, such as the following for the 340th, helped to heighten morale and instill justifiable pride in the men of this bomb group. This was another thread that the ever-alert Heller wove through *Catch-22* as Col. Cathcart tried to court the *Saturday Evening Post*.

The 340th Bests Cyclones, Vesuvius, and Hitler
By Gill Robe Wilson
New York Herald Tribune

NOVEMBER 13, 1944

One of the toughest, happiest outfits I have ever hitchhiked a ride from is the 340th Bombardment Group, commanded by Colonel Willis Chapman. They have no doubt but that they are the best outfit on earth and after studying some of the targets they go after, it is possible they would reach the finals in an elimination match. They fly Mitchells, which they assert are the direct linear descendants of the sweet chariot.

Up to here the same idea prevails in every group but from here on the story is entirely different, for the 340th rejoices in the name of the "Unlucky 340th." What this outfit has survived and surmounted has become the foundation of such a confidence as amounts to fanaticism. Black cats, graveyards, three on a match and such normal tokens of ill omen are powerless with the 340th

CYCLONE STRIKES TWICE

Back in the dark ages, just after Pearl Harbor, this outfit was formed for training. Just as they were going well, a cyclone struck their field and stripped them of airplanes. They moved from the wreckage to set up at another base and the same thing happened again. Finally re-equipped, they started for the war and hit the worst weather the South Atlantic had produced in years. Some were lost but most of them straggled into Africa and finally made their bedraggled way to Montgomery's army. When finally they got set for action and went on their first raid, the Germans were waiting for them as they do for every new outfit.

The group commander, navigation officer, bombing officer and one squadron commander were lost on the first raid. The group received a new top echelon and dug into its job. From Tunisia up through Pantelleria and Sicily into Italy they bombed and strafed with fierce determination to lick their jinx. They ranged into Greece and Albania, Yugoslavia and Bulgaria making accuracy a fetish. Then another commanding officer was shot down to become a prisoner of war.

One of their missions from Tunisia had been to bomb a famous country club sort of airport in Sicily. It had a swimming pool and luxurious hanger and crew quarters. Gleefully they took the place apart and drove the Germans out. Next week they were ordered forward to occupy the field themselves and had to rebuild what they had destroyed. Cussing their own accuracy they put the place in order and immediately afterward were sent forward into Italy.

Here they took on a new type of work with strange instruments and different procedures. Just as the 340th was getting good at the fresh assignment, Vesuvius erupted and cleared the works. Ashes and brimstone [covered] airplanes, tents, equipment and trucks. The squadron escaped with the shirts on their backs and nothing else. But were they licked? Not they!

Four days later from another base and with borrowed airplanes, the "Unlucky 340th" was out against the Germans in full strength. They went through the long months before Cassino and when that show came off put every bomb into the target area. From Cassino to Rome they hammered bridges and transport and finally moved to Corsica to participate in the coming invasion of southern France. The night they moved to Corsica the Luftwaffe made its one effective counter-attack and, of course, the 340th had to be underneath when it came off.

Once again they re-outfitted and went back at the Nazis without losing more than a day of combat. Acting on a long target mission, their luck started to change and the gods smiled very broadly. Many of the group made dead-stick landings at home base, so narrow had been the gas margin. But no ship was lost

LUCK CHANGES AT LAST

Since that day the 340th has been hot as a firecracker and nothing seems able to go wrong. And, like good polo players, the group is pushing its newfound fortune. Bombardiers are racking up hits with monotonous regularity and nobody around the place seems the least bit war-weary.

But there is something behind this conquest of fate and newfound smiling fortune that is very solid. It is the percentage of sweat. I saw a dozen homemade gadgets which the crewmen were busy perfecting to improve bombing, shooting and navigation. They are never content. The charts show they pass two hours practicing for every three of actual mission time. The chaplain and flight surgeon are like rooters at the big game. Colonel Chapman is the coach and everybody gets in the game at one time or other.

With sweat and tears and the sand of character for mortar, out of the stepping-stones of misfortune the 340th has built itself a great foundation. Nothing can happen to it anymore—both Vesuvius and Hitler have become incidental."[3]

To the men of the 340th's great credit, in thirty days thirty missions were flown over the deadly, heavily defended and prepared Brenner Pass and thirty successful strikes were made. The Germans, fully cognizant of Allied intention, had vainly tried to defend this twisting and turning supply line by moving in considerable anti-aircraft batteries. As mute evidence of the effectiveness of these medium bombers, these B-25s manned by their increasingly skilled and determined crews, there was not a single day in February 1945 that the Brenner Pass line remained unbroken. The unbelievable accuracy of this daily aerial pounding continued. The numbers soared as seventy-seven consecutive hits were made on targets. These targets were bridges, the most difficult of all targets to hit, and all were destroyed with pinpoint accuracy, thus earning the outfit the nickname of the "Bridge Busters." This group set a world's record for bombing accuracy when it scored better than 90 percent accuracy over a 3-month period. As well, it destroyed more bridges with less tonnage than any other bomber outfit.

The "Unlucky 340th" Bomb Group had risen, like prized cream in milk, to the top.

The Best Damned Bomb Group There Is!
Yes, damn, indeed!

It was 1984 and the 340th members were gathering for one of their highly anticipated annual reunion banquets. The room, prepared for its expected good turnout, was large, and the tables, covered with white linen cloths and fresh, brightly hued flowers, marked the importance of this day. Members, maintaining the formality and manners of earlier times, were continually arriving, men in suits, and women, usually, in glamorous cocktail dresses. The steadily increasing hum and buzz of voices, punctuated with hearty exclamations of joyful recognition, permeated and echoed about this spacious, inviting gathering place. The pleasure was tangible.

This year George was Master of Ceremonies. His talk from the wooden podium centered on the four commanders of the 340th Bomb Group and, primarily, on one man who, with Charlotte, his elegant, gracious wife and sweetheart from age 15, was also in attendance.

George L. Wells as Master of Ceremonies during an annual 57th Bomb Wing Reunion.

SELECTED TEXT FROM GEORGE'S SPEECH

"Then came Chapman, commander number four. Mr. 340th himself. I knew we had a new commander because he hadn't even unpacked his B-4 bag, when on the night of March 21st, he must have lit the fuse to get our attention—boy, did he know how to get our attention—the whole top blew off Mt. Vesuvius. With that blast Chapman and the 340th really went into the history books: the only combat unit in history to be put out of action by a volcano.

"He instilled pride of unit in us," George continued later. "He taught us how to do the job and led us on to greatness with loyalty and patriotism. Our Len Kauffman, 489th Squadron Commander once said to me, 'It isn't for a junior to rate his commander, but on a scale of 0-10, Bill Chapman gets a top 10.' All of us agree with Len."

"In addition, he did something that has had a lifetime effect on my life:

the most important thing that ever happened to me. I don't think he knew he had a part in it, but I'll tell him tonight.

"In the middle of May 1945, just before I was scheduled to return to the group after my 45 days rest in the States, he sent me a message saying, "The group's being returned to the States for redeployment in the Pacific. Don't let them send you back. I'll pick you up when we hit the States." At the Atlantic City Redistribution Center I ran into Dr. Nestor from the 489th. He solved my problem by prescribing 30 more days of rest for me to await the return of the Group. The Group never went to the Pacific, but during those extra 30 days I met a beautiful young girl by the name of Shirley whom I later married—just a little thing, as that message from him charted the rest of my life."

Willis F. "Bill" Chapman and his wife, Charlotte.

George concluded:

"You all have known certain bosses, commanders, or leaders who stood out to you. Yes, Bill Chapman, even though I originally resented you for your long cigarette holder, your three- to five-minute straight-and-level bomb runs, and the denial of the Group's stay at Capri after Vesuvius erupted, you quickly changed my mind. You were and always will be 'My Colonel.'

"I think these words apply to Bill. I understand they are inscribed in bronze in the Library at West Point, where he may have seen them while he was a cadet:

"Let us remember that 'tho the paths of glory lead but to the grave,' if when the sun lies low in the West, and the light grows dim, we can look calmly back over the years and see a path straight and clean, unscarred by acts of weakness or dishonor, a path along which the only visible evidence is a fight well fought, a faith well kept, and a race well run; then we may know that by the years of our service our life has justified itself, that through keeping faith with our dead we have kept faith with ourselves, with our corps, with our country, and with our God."

"Let's look in your West Point trunk, Dad," I said. I am always drawn to this trunk and it had been many years.

Upon graduation every cadet left The Point with one of these large black and brass wardrobe trunks designed for military transport vessels. My father's class of '35 was no exception. There was a place for everything. It offered multiple openings that gave access to the hanging wardrobe of military jackets and pants hanging on wooden hangers. There was storage for riding boots and spurs and sabers, shelves for caps, small drawers for miscellaneous small items.

I placed his tall, black-plumed "Tar Bucket" hat, used for formal dress and parades, over my long hair. The formal jacket to be worn with this hat displayed impressive large, round brass buttons embossed with USMA. The informal jacket had no brass buttons and was worn daily paired with a hat bearing the West Point insignia.

"What is this white fabric?"

"Parachute silk," he replies.

There is his Sam Brown belt with brass buckle; his watch; an issue canteen; W.P. wool blanket; patches; buckles; ashtrays and lighters; a large, folded and brightly painted piece of an Italian airplane skin; the bomber jacket for warmth, and the long, green canvas one for milder temperatures.

And here, my favorite — custom-made, richly smooth and pliable British Peale leather riding boots. Drool-worthy (but too large) for a daughter whose totem since forever has been the horse.

Long ago I was confirming that, of course, his favorite activity was riding. "Not exactly," was the reply. "I never saw a horse I wouldn't just as soon shoot."

Yes Indeed stood a formidable 18 hands and my 6'2" father had to drop the stirrup leathers to the longest length just to mount, then shorten them from atop, much like pulling up the ladder after oneself. "I don't think that damned horse ever walked a step in his life. Just jiggled and joggled and pressed the horse in front. Every once in a while that horse would give him a couple of hooves. And that always stirred the pot up."

The yearly cavalry field trip was not anticipated, and Grizzly raised his head. "Grizzly was not as big but also never took a walking step. We were out five days and it rained four." Sadly, here was a candidate for the second chamber.

Don't get him started about his mount for the jumping ring.

He would start running and I couldn't get him to quit. I had gone through the hurdles all right and tried to get back in line but the horse had his own mind and went through the jumps again. I thought, I can't keep going through this damn thing again, and besides, my tail was getting sore. So I figured I'd swing wide around and aim for the rumps of the waiting horses. He wasn't that big, not like Yes Indeed, but he was just a mean bastard so I hung on for dear life. He put all four feet down and I put in a flying block into the rump of that front horse. I almost knocked the horse down. Funny thing was, it bounced me back on my feet so I didn't fall down on the ground. It stood me on my feet and, if I could have kicked high enough, I would have fixed him up too.

"Dad, are you sure this is parachute silk?" I asked, returning to the white fabric. I gently began lifting the silky fabric from the drawer. As tiny covered buttons were appearing, then a bit of lace, I focused in on what slipped out to reveal my statuesque mother's very slim, bias-cut, Queen Anne collared, satin and very elegant, long-trained wedding gown . . . rising like the phoenix from its unbefitting dwelling. Or, on reflection, perhaps that was exactly where it belonged.

AFTER THE WAR

Bill was a career Air Force officer, which included commanding the USAF's first jet bomb wing and being instrumental in the success of the British Harrier Vertical Takeoff aircraft. He retired as a Brigadier General and become Ling-Temco-Vought's aerospace director in Europe. He held a patent for dealing with oceanic oil spills.

At Bill's funeral service at Arlington National Cemetery, this author's son, Jason, gave the difficult eulogy that well described his grandfather. A segment follows:

> Granddad was an eternal learner, an eternal student, a curious man with broad interests, but he was also a practical man who could not only see the big picture but lived it as well. Through Granddad, I grew to be of a spirit that anything is possible. If I weren't to be a general like him, I'd be something equivalent. No dream would be unobtainable. He was always there for everyone. Even in Granddad's last days, he maintained his unique wit, and he never complained about anything.

> Granddad was a best friend of mine, and he was a hero. Optimism, curiosity, the passion for a cause, and motivation to "turn the blinking crank" and do something about it—these are the greatest gifts that Granddad gave to me, and, no doubt, gifts he shared with all of us.

CHAPTER 5

BOB: FROM NEIGHBORS ORVILE AND WILBUR, TO *CATCH-22*'S GENERAL DREEDLE

" **O**rville signed my first pilot's license (#185), years late. But, when they were in Auburn, it seemed like all they talked about was their bicycle shop in Dayton, Ohio."

Aviation in its infancy came to Auburn, Alabama while young Bob Knapp was not yet in his teens. His family lived three houses away from, and his mother helped, a lady who ran a boarding house, and it was here that Orville and Wilber Wright lived for a few months while trying to perfect their budding airplane invention. The brothers became acquainted with the Knapp family as they conferred quietly and privately with a Professor Fullen of then-Alabama Polytechnic Institute (now Auburn University) about adapting their airplane to military specifications. Primarily, they needed to figure how to quickly disassemble their airplane to allow it to be transported on a two-horse wagon. The Army would purchase the plane if this could be accomplished.

These two young men were cautious in their conversations. They did not want the people of Auburn to know their visit had anything to do with an "airplane." They had been the brunt of much rib-poking, eye-rolling, and flippant jokes about their dubious states of mind in trying to develop something huge that they intended to put in the sky. They requested of Professor Fullen and Mrs. Knapp to say that they were bicycle shop owners and were working on a new type of bicycle.

At this very susceptible age, young Bob was afforded rare exposure to alien talk of airplanes and flying. The genesis, the heartbeat, the very birth of what would become an all-consuming passion began here. The seed was planted and continually watered and nurtured over the course of his entire life.

Bob had yet to finish high school when World War I broke out and he, with exuberance, immediately hustled to enlist in the military. Because of his youth he was rejected, but he discovered that the Army Cadets would welcome him. He signed on and then returned home a bit apprehensive about telling his mother of his plans. The surprise was his.

"I thought she would be unhappy about it. In those days people's parents had a lot to say about what they'd do until they were 21 years old.

"I told Mama I'd joined the aviation section of the Signal Corps and was going to learn how to fly airplanes, and she said, 'I think that is just fine because Orville and Wilbur tell me the airplane has a great future in it and maybe you will get in something good. But always, I want you to just fly low and slow, Robert.'"

During WWI, at age eighteen, Bob enlisted, serving in England. While neither he nor his unit, the 92nd Bombardment Squadron, saw combat because the propellers ordered for his unit's Handley Page 0/400 bombers failed to arrive in time, his ability in, and passion for, flying steadily increased.

While still an enlisted man, he served in a detachment of flying cadets. He described those early planes as being four-cylinder jobs totaling 8 horsepower, water-cooled but with few instruments: a red-line thermometer sitting on top of the radiator, a small compass, an air speed indicator, a tachometer, and an oil pressure gauge. After four hours of instruction you were expected to solo. His flying instructor quickly gave his blessing for Bob to solo before he killed either one or the both of them.

Shortly thereafter Bob learned all about night flying and bombing. "Night bombing was kind of interesting, but not very accurate. Our bombsights consisted of a couple of nails on the side of the fuselage. Even with hours of practice, we seldom hit anything, and when we did it was ninety percent luck."

The war's ending found Bob, in 1919, then flying border patrol against Mexican guerrillas, including the notorious villain/hero general Pancho Villa. In airplanes armed with machine guns, "we kept them south of the border," he confided with a slight grin. Man and machine, at that time, were each lucky to last the three hours between posts. Emergency landings were standard, either to service the airplane or to allow the pilots to satisfy nature's calls. They took the dependable, grey and white plumed homing pigeon with them. Were they to have an accident or be wounded, the bird was released to fly back to base and spread the alarm.

On one patrol he paired with his flying buddy and friend, Bruce Struthers.

I flew to Douglas, Arizona with Bruce. There we met his father, an engineer with Southern Pacific Railroad. His dad revealed to us that he was about to pilot a train across Texas to San Antonio. This gave Bruce an idea.

"Dad, I'll fly by you and give you a buzz just about the time you get to Alpine."

The idea pleased his father. "I'll be looking for you, son."

The next day Bruce took off to catch his dad's train at Alpine. In that area of Texas the terrain is irregular. The hills rise like cones, almost volcanic in nature. As a pilot, if you were in the least bit careless you could have serious problems.

Bruce Struthers, true to his promise, flew alongside his dad's train, watching his father and waving to him. His dad blew the train whistle a few times and waved proudly at his handsome son.

And then it happened. His father watched in horror as his son, Bruce, flew directly into one of those deceiving Texas hills. To this day I still remember the tragedy of Bruce Struthers.

The Mexican border has always played a significant role in the growth of North America. The Border Patrol is a part of that history. During its first year or two the accident rate of aircraft flying patrol was so great that it became the rule rather than the exception. The reasons were varied but each was major: novice pilots; unsuitable (because it was too thin) Liberty oil; boredom, creating reckless pilots; parachutes not available; extreme heat during the day and debilitating deep-freezing midwinter nights.

When the Border Patrol moved to Nogales, Arizona in 1921, Bob was given command for two years patrolling the Arizona border. He had known that if parachutes had been available, some of the fatalities could have been avoided and he was now in a position to try to correct this.

In his words:

I had sent my supply officer, a fellow named Wolfe, up to Dayton, Ohio to learn how to pack parachutes. He got a supply of them and brought them back to us, so here we were with new parachutes and Wolfe was going to show us how to use them. Of course no one had taught us how to jump with them. Wolfe showed us how to pack them and fix them.

Finally he decided he was going to make a jump. We notified the newspaper and they put in the paper that Wolfe was going to make a parachute jump. People had never seen a jump before, as a matter of fact, until we came, and airplanes were a mighty scarce thing in that area. Everybody that had means of transportation came out to the field to see this parachute jump.

I flew Wolfe for the jump but just before taking off, Capt. Usher said, "Bob, let's take a cat up and drop him in one of those little pilot chutes." The little pilot chute was on a spring that would snap open and pull out the main chute from your pack, but it wasn't much bigger than a large handkerchief. It was about the right size to let this little kitten down, so we talked it over. It would be a good idea to do something unusual, so we made a harness for the kitten. We tied the parachute onto the kitten and rolled the cord around the spring. We gave it to Wolfe and told him the first time over, let the kitten go. Then we will fly around again to see where the kitten lands so we can tell more about the wind and judge better where you are going to land.

I flew around and told Wolfe to release the kitten. Instead of the boy unwinding the string around the kitten and chute so it could open, I guess he was excited, so he just dumped the whole thing over and the parachute had no way to open up. The poor little old kitten fell down and used up all his nine lives right there. The cat was finished for sure, but the funny part, if you can call it funny, people didn't know about the size of the parachute and they thought it was Wolfe. Women fainted but I didn't know that up in the airplane. I flew around and dropped Wolfe, of course, he came down and made it all right. Everything was OK but a lot of people didn't speak to us for a long time for what we had done to the poor little old cat. The kitten would have been all right but Wolfe got excited and he was a little scared. Nobody had ever seen anyone jump out of an airplane with a parachute before, I never had—none of us ever had, it was the first time. It was just too much for him to think about the kitten—he just tossed it over the side just like it was a brick, it was too bad, but you just couldn't do anything about it.[1]

With other pilots, Bob barnstormed across the country trying to arouse America's interest in aviation. He happily tried everything he could: air races, parachute jumps, flying airmail routes.

Much of his life from 1927–37 entailed flight training of the Army Air Corps. Bob Knapp said, in fact, that he had trained most of World War II's Air Corps officers.

"Military life came very easy to me. I enjoyed every bit of it."

His leadership qualities soon became apparent. Robert Knapp was a natural leader, who was highly respected by the men under him. Having been born in 1897, this day-after-Christmas baby was becoming an accomplished pilot before most of the men who served under him had even been born.

Bob tromped confidently through life learning by the seat of his pants, for there was no airplane history to draw on. "When the Japanese attacked Pearl Harbor I was playing golf with some of the Squadron COs at the Jackson (AFB, Miss.) Country Club. We thought it a wild rumor but called an Officer's meeting anyway. Confirmation was soon received from Washington and the Sunday of Dec. 7, 1941, we were instructed to move and we did."

Bob entered WWII with his experience and leadership qualities expanding, and he eventually pinned on the coveted Silver Star of a Brigadier General.

Heller shrugged and yanked Knapp's chain as equally as he had done for everyone else with his creation of the factious General Dreedle.

General Dreedle drank a great deal. His moods were arbitrary and unpredictable. "War is hell," he declared frequently, drunk or sober, and he really meant it, although that did not prevent him from making a good living out of it or from taking his son-in-law into the business with him, even though the two bickered constantly. *(p. 212)* [Gen. Knapp's son was his Aide de Camp.]

Being on the ground floor of the flight experience, coupled with longevity, afforded Bob Knapp a wealth of unique tales to tell. Story after interesting story piled up in his wake.

On two occasions he crossed paths with the famous ski-nosed entertainer and comedian Bob Hope and his USO Show. For several hours the fighting troops were distracted and entertained by much-needed and appreciated jokes, singing, skits, and laughter. Hope was always a favorite; he unfailingly brought with him talented, and often beautiful, stars for these highly anticipated diversions.

The first encounter happened in Tunis in early April 1943. Hope and cast visited the 321st. It was an early Sunday morning. Abruptly the Squadron Mess Officer burst into the cook shack with news that Colonel Robert D. Knapp had selected the 445th to feed Bob Hope and his USO troupe at lunchtime. The cook shack was a Quonset hut with nothing but a dirt floor.

The mess crew worked all morning preparing a thousand meatballs and barrels of spaghetti for Bob Hope's team and the men of the 445th. Lunchtime came, but Bob Hope didn't. Since the men had to be fed, luncheon was served. Not unexpectedly, the 445th found the food just too good to be wasted, so they consumed everything!

About 1:15 p.m. the Mess Officer returned with word that Colonel Knapp and the Bob Hope troupe were on the way. Out came cans of boned chicken and boxes of rice. All this was mixed together in a ten-gallon roasting pan. Soon it was piping hot and ready to be served, and then it happened. The two men carrying the roasting pan across the room dropped everything on that dirt floor. There was, of course, only one thing to do. The two men scraped up the chicken and rice and quite a bit of dirt, and served it all, attractively, on the best china available.

An hour later Bob Hope dropped into the cook shack to visit the mess staff.

"Men, that was the best feed I have had on the tour so far! Thanks very much."[2]

Months later, on Corsica, Knapp again welcomed Bob Hope, the lovely Frances Langford, and Jerry Colonna, Hope's zany sidekick, when they were invited to put on their show for the troops there. Afterwards, they were escorted around the area. Upon cresting a sand dune to view the splendor of the Mediterranean, they were treated to the glory of a beach filled with our totally naked American soldiers who were, in seal-like manner, swimming and sunning themselves. Frances let out a yelp, did a 180, and all returned to the mess tent.

While Bob Knapp's stories were abundant, informative, and often humorous, some had to include a darker side. These usually involved uncommon acts of bravery. From what he personally witnessed in battle as well as acts that were brought to his attention as being extraordinary, he was inclined to agree with William Manchester who stated in his book Goodbye, Darkness, "As men, I know, do not fight for flag or country, for the Marine Corps, or glory, or any other abstraction—they fight for one another."

"I really believe that most acts of personal bravery on a battlefield occur because the man is really thinking about those buddies of his and that he is helping them," agrees Bob Knapp.[3]

The following story of a bombardier epitomizes this. This particular plane got into severe trouble with punishing flak over the target and the crew was having to abandon ship. The bombardier had left his parachute in the crawlway, as was common, so he could more easily inch his way into the airplane's nose where he would have more room to maneuver without it. When the pilot gave the order to abandon the dying ship, the bombardier retreated back to midship, pushing his parachute ahead of him. It fell onto the floor—the floor that was no longer there because the engineer had released the escape hatch. His parachute, his lifeline, dropped out.

The pilot ordered, "The rest of you chaps abandon the ship now. I'll stay and bring this airplane to a landing of some sort."

The bombardier said, "No, you can't do that—this thing is too badly shot up."

Knowing that the pilot was not going to leave him, this man, this bombardier, without a parachute, then jumped out of the airplane.[4]

While his formal education was never completed, still Bob Knapp continued to advance. As his flying ability and leadership skills were recognized he rubbed elbows with some of that era's military elite. General John J. "Black Jack" Pershing, General William "Billy" Michell, General Henry "Hap" Arnold, General Carl A. Spaatz, General Claire Chennault, General Ira Eaker, General James "Jimmy" Doolittle, fighter ace Captain Edward "Eddie" Rickenbacker were a few of his working acquaintances. It was, in fact, Bob Knapp who had trained the crews that flew, for the first time ever, their B-25s off the deck of the carrier *Hornet* in the historic and celebrated "Jimmy Doolittle Raid" over Toyko.

> He had no taste for sham, tact or pretension, and his credo as a professional soldier was unified and concise: he believed that the young men who took orders from him should be willing to give up their lives for the ideals, aspirations and idiosyncrasies of the old men he took orders from. *(p. 214)*

Bob Knapp flew 50 missions and retired with 14,000 hours of flying time, more than any other service man at that point. While he was never shot down in combat, Bob did admit to several crashes. "Crashing is no big deal," he said in 1991. "It's the getting up that's the hard part. I've totaled a lot of them in my day. I don't know how I walked away from some. But I'd like to do it all over again."

"I did what I wanted to do. I just loved to fly, I really did. I'd go anywhere. I always felt like I was big enough to do what I wanted to do."

The years passed, and on the cusp of his 90th birthday, Bob visited Wright-Patterson AFB, Ohio, for a dedication ceremony of a B-25 Memorial for the 57th Bomb Wing and a reunion of members of his unit. Lt. James Nylund, Special Activities Officer for the AF Institute of Technology, was assigned as his escort officer. With his first glimpse of Gen. Knapp, as he was disembarking, Nylund immediately realized that this was, indeed, an uncommon individual. Here was not a cobweb-chewing, aging man drifting into his twilight years. This man was expressing his thanks to the crew for the flight with a gleam in his eye that said "flying is everything in the world." Ignoring protocol, Bob, with his walking stick, settled in the front seat with Lt. Nylund so that they could talk more easily as they headed for the Visiting Officers Quarters.

The flight had been delayed and it was the deep-dark hour before midnight when their yellow-white headlights swept through the black and across the very quiet hotel's facade. Bob asked if the lieutenant could do him a favor after he got checked into his room. Members from his old unit were staying there and he'd like to see "his boys" before he turned in for the night.

Nylund did not think, at that late hour, that many of "his boys" would be around, but, to his surprise, there were at least 40 of "his boys" eagerly awaiting his arrival.

"General Bob, we were hoping you'd make it in tonight."

"General Bob, it's really great to see you."

"Sir, you look great; glad you could make it."

He greeted each by their first name as they shook hands. Admiration beamed from the men of the 57th Bomb Wing. Pride beamed from the General for the men that had served in his unit.

"Many of them thanked me for bringing him and asked that I take good care of their leader—if I could keep up with him."[5]

AFTER THE WAR

After the war, Bob became Chief of the USAF Mission to Argentina. In 1953, after a 36-year career in the USAF, he retired to his farm in Auburn, Alabama. He became a founding member of the Order of the Daedalians (a fraternal organization of military pilots) and remained extremely active with reunions of the B-25 units he once led.

He had been awarded The Silver Star in 1943

For gallantry in action. On March 31, 1943, Colonel Knapp took off on a sea search mission leading fourteen B-25 airplanes. The weather was bad with rainsqualls and poor visibility to a point about twenty miles out to sea. The fighter escort and six of the B-25 bombers became separated in the bad weather and returned to base. Colonel Knapp gallantly continued the search with the remaining eight bombers. At 12:55 a convoy of six ships, two of them large with fighter and marine escort, was sighted. Colonel Knapp's formation was attacked by fourteen enemy fighter aircraft, and although the tail of Colonel Knapp's lead aircraft was damaged by two explosive shells and machine gun bullets, he gallantly and skillfully led the formation in destroying one enemy aircraft, damaging four, and losing the others in the clouds. Colonel Knapp then gallantly led the formation back to the convoy, climbed to forty-five hundred feet, and made the bomb run, sinking one large ship and seriously damaged another. By his gallantry and devotion to duty on this occasion, Colonel Knapp has upheld the highest traditions of the Army Air Forces.

George L. Wells

Capt. George L. Wells

Joseph Heller

Lt. Joseph Heller

Gen. Robert D. Knapp

Col. Willis F. Chapman

Bob

22X 29-361 2ND LT W.F. CHAPMAN

Willis F. Chapman

In his novel, Joe Heller had a good time with those four men on the opposing page - and a slew of others. All of them were in the same bucket as each was dissected and refashioned. Humor helps greatly to make civilizations work and, when it serves up an insightful message as well, it becomes a powerhouse.

The men of the 340th Bomb Group were blindsided when *Catch-22* skyrocketed into being. The nose-out-of-joint reaction, the what's-wrong with-that-guy knee-jerk response eventually cooled enough for them to inspect the book for what it was - an original, wickedly funny, shudderingly insightful, masterpiece of literature. Then was when personal chafing actually developed a tongue-in-cheek-pleasure in being part of The Great *Catch-22*.

"I'd like to see the government get out of war altogether and leave the whole field to private industry" - Milo *(p254)*

CHAPTER 6

GEORGE, AGAIN

It has been seventy-four years since the sounds of WWII ceased, when the length of a man's life was in question as he arose each morning. The conclusion of the war allowed a return to home, family, and the familiar way of life. Although these lives were now freed to pursue chosen directions, still, decades later, memories of the times of conflict remain deeply embedded.

George's post-war path has been full. He became a fighter pilot based in Alaska, then a commander of a fighter squadron, retiring as a full colonel. He was vice-president of one company and president of another. But nothing has dulled the memories of his combat experiences.

The events and effects of WWII are still vivid in my mind. It isn't like everyday living in peacetime, where you only remember the highlights and low points of life. Combat leaves a deep lasting memory. I turned 91 in Jan. 2010 and my memory isn't good but that doesn't apply to the biggest war.

I remember the beginning as I stepped away from American soil. I was at Columbia, South Carolina Army Air Base, as pilot-commander of a B-25. The commander assembled the members of my crew in a rather random fashion. In this case, three officers filled the positions of pilot, copilot and bombardier, and three enlisted men filled the positions of engineer, gunner and radio-gunner. Anyway, as it came time to board a truck that would transport us to the aircraft that was [the first step in going] overseas, I waited with two other single crewmembers while the three married ones said a tearful farewell to their wives. As we all started to walk toward the plane, the officer who was the bombardier caught up and walked quietly beside me for a few minutes. Suddenly he turned to me and said, "I'm never coming back." I immediately said to him in a rather shocked and tongue-tied voice, "How can you say such a thing! We're not even there yet. What makes you think you're not coming . . . I'm not thinking that! I'm intending to come back . . . that's not gonna . . . I'm not gonna . . . sure . . . it MAY happen to me, but I'm

NOT going over there thinking its gonna happen to me, and every moment I'm there I'm not gonna think its gonna be happening. You gotta take that idea out of your mind!"

This guy's name was Reichard. They called him "Red" Reichard. Now, even though the six of us went over there as a crew, our squadron chose to continually rotate the six positions, so you never knew who was assigned to fly any particular mission with you. Now, this guy was such a pessimist, convinced he was NOT gonna make it. Well, when we got over there as a unit, we knew each other, so when we bunked down for that first night, we slept near each other on the ground. We chose to stay near each other even though later on we wouldn't necessarily fly together. Anyway, we kept sleeping near each other, and every time this fellow Reichard saw his name up on the board that he was going to fly the next day, he couldn't sleep. He rolled and tossed, and just couldn't sleep soundly the whole night long. Imagine that! Just tossed and turned. But, in spite of his fright, this man went on to become a lead bombardier in our squadron. Can you imagine that? Now, this really speaks something well. This guy is scared to death and thinks he is going to die, and yet he gets SO good that he is selected as the lead bombardier based on his performance. That's a real hero. For myself, I never had any fright, and I could do the job real, real well. But Reichard could do the job real well plus with his fright. I say when you look at heroes, and I may have some great medals for things I did, but I did nothing like that guy did.

On one particular day, during a rather big invasion, the 488th sent B-25s as the first bombers in the air as troops were attacking on the ground, and our battleships fired from behind us. Now, the Germans had fighters up there too, and they could get to you, but of course, as bombers, the thing we had to worry about most was the anti-aircraft artillery coming up at us. During that time period, the Germans used radar to determine the approximate altitude of our planes. Their gunners on the ground used that information to adjust their settings based on estimated flight speed. They used an 88 shell, which would explode at a predetermined altitude, and anything nearby at that moment would get showered with shrapnel or flak. But those Germans weren't stupid when they designed these shells, they also had a contact fuse on the nose, so if the shell made contact with anything, it exploded through that second method.

Well, as it turned out, my squadron was the first one in that attack while several other squadrons filled in the rear of the formation. Reichard, being the lead bombardier, was in the lead position of our box of six planes. He gallantly went in on that target as my plane followed maybe 100 yards behind. The flak was coming up quite thick, and it was pretty bad.

Suddenly, the Germans' anti-aircraft artillery seemed to zero in on Reichard's plane. I have this mental picture of his plane right after the shell exploded, and it's completely on fire. It was really awful and left me with a rather helpless feeling. As Reichard's plane went down in flames, not a one of them got out. This is a vision I will never forget.

Actually, it's difficult to determine whether the explosion in Reichard's plane was from immediate contact of being hit by a shell, or just the time in the air while being hit with an onslaught of flak, because everything on the airplane was a mess. Now, that contact theory is the simplest thing there is, but after I landed my airplane at the end of that day, there was a hole going in from one side of my plane and out the other. The contact fuse didn't work! And there's the testament that "I'm not coming back" could possibly be a self-prophesizing statement. In contrast, I said, "I'm coming back, and I'm not going to think anything different." Every one of a B-25 crew was constantly in danger while over enemy territory, but while in combat . . . one MUST stay positive.[1]

See mission #37

George, from his plane, had helplessly witnessed 8H, the ship of his friend, Lt. "Red" Reichard, spiral down in flames. After the war George himself returned Richard Reichard's effects to his widow.

As a pilot, you have a feeling of controlling your destiny to some degree, whereas the other members of the crew have to rely on you. I've always believed that the real heroes in the air war are those brave gunners, bombardiers, engineers or navigators doing their job while being shot at and wondering how serious the situation is to their survival. I really felt a great responsibility for the crew members. I wanted to do my job in such a way that they would want to be on my plane when they flew a mission.

George's plane with hole in its side

Composite of Major Garski, Lieut. Dean, and Lieut. Reichard. Dean's ship, hit by anti-aircraft fire, bursts into flame, goes down out of control, crashes with its full load of bombs and explodes (lower photo). Bombardier/navigator, Hawley, parachuted to safety—in a German prisoner-of-war camp.

The end of the war was imminent in April of 1945 and the 340th was transferred to the Adriatic coast of Italy, near Rimini. The men of the 340th took with them warm feelings for that lovely island of Corsica and its gracious people of whom they had grown quite fond. Hal Lynch, a bombardier from the 488th squadron recalled:

On a clear Sunday morning in May, just one month after the move to Rimini, handsome Len Kaufmann, 489th commander, flew a B-25 back to Corsica just for old time's sake. He found the 340th runway to be a walking path for a herd of cows. Len buzzed the runway a couple of times to drive the cows away. As he approached the runway he spotted a woman and two children running out to greet him. This had to be Rose and her two sons, Guy and Ange, a family who lived on the edge of the 489th site. These people had recognized the B-25 and its markings and had run over a mile to the former 340th runway. For Len Kaufmann that reunion in front of the B-25 with those three friends will always be a heart-warming memory.

In late April 1945 the 340th completed its final combat mission over enemy territory. Flying out of Rimini, Italy on the beautiful Adriatic, the 340th saw the once-powerful German Army in full retreat in northern Italy following over two years of combat duty that had commenced in North Africa. The war in Europe was soon to end.

Following the laws of the universe the years whisper by, each one taking more of these aging WWII warriors along with it: General Knapp, Colonel (later General) Chapman, Lieutenant Heller . . .

At yet another more recent reunion of those remaining and still able, this author and her husband, Jim, joined George and Shirley Wells' table. World War II was thick in the air as stories and friendships intertwined. The room buzzed with the sounds of husky voices as these men met once again. Former pilots could always be singled out. These were the men who kept gliding to the left ensuring that their "good ear"—their right ear, the ear that had not been impaired so long ago by the B-25's screamingly powerful engines whose propeller tips were only 8"–10" from their left ears—could be turned in the most advantageous way to hear the conversations. This subtle shifting about was quiet and constant with the men continually ribbing each other about their common affliction. Friends, family members, even an occasional reporter mingled in this crowd. Soon the remembrances, the stories, the glue-binding experiences began to filter through.

"And with hands held to represent two banking airplanes—old fliers really do talk this way—one of them said, "An inside turn is okay if you gentle into it, but Henry, he used to cut it sharp . . . you'd have to push your throttle into the wall to avoid running into him." The hands turn edges down toward the floor in a gesture of banking and then suddenly lash out in closed fists in a dramatic illustration of pushing a throttle to the wall".

"Say, did you know that Dita Beard of ITT fame was a Red Cross girl with the 489th Squadron?" asks Ned Heilwig, who then was a dashing, mustached wing-pilot. "She used to be meeting us with the coffee and doughnuts when we returned from missions. Why, I used to dance with her," he recalls. "She was a real great girl . . . used to swear like a trooper."

"Pantelleria was an island fortress, the only one in history to surrender to the American Air Force after a bombing. Col. Chapman flew down alone to pick up the surrender papers. Talk about nerve, they say. Talk about daring."[2]

Friends here and those no longer here were all fully present. As the evening began to quiet and the festivities shifted to lower gears, these four found a quiet corner so Pat and George could talk shop.

Pat:	George, tell me details. How did you dress for flight?
George:	Proper equipment was taken in case you had to bail out. Clothing-wise, tent. I always wore a parachute in the plane.
Pat:	How did you enter that plane?
George:	There were ladders on both sides. You got in the seat and pulled the strap across your lap. You would pull it off to bail out. A key was used to start. There were two pedals: brake and accelerator. The cockpit had good metal down the lower side for protection. Plexiglas, for visibility, was from top down to the start of this side metal.
Pat:	How did you start the engine?
George:	The ground crew started it. Mechanics pulled the propellers.
Pat:	How many and who were in the airplane?
George:	In the lead ship there was the pilot, copilot, navigator, bombardier, three gunners (top, waist/radio operator, and tail).
Pat:	How does the plane progress down the runway?
George:	It takes about fifty times the length of the plane to get in the air. It would start slow and build up speed. The B-25 was a powerful plane.
Pat:	What did it feel like?
George:	It was a little shaky but it was not a problem. In the air it settled down very good. It needed full power to reach elevation. Then cut back. Noisy—was the way it was built. The propellers were right by our ears and we had no ear protection. We were supposed to put cotton in our ears but we didn't. In the pilot area, we were the only ones near the propellers, which is why most of us pilots now have a hearing loss, especially on the left side.
Pat:	Yes, I am always aware of you men continually moving around each other to place voices on your "good side." How was the crew chosen?
George:	Airplane crew could be mixed. It could all be from one squadron or mixed with other squadrons. The leader could fly in any squadron. Pilots did not always have the same plane. We flew whatever was ready. The pilot's battle is up high inside the aircraft. We dropped bombs and got out. On a bombing run, as the leader I would roll right, then left, up, then down

for evasion until I got within about half a mile of the target. Then we would go straight to hit the target. This was the most dangerous time. We'd drop our bombs and immediately leave the target. Sometimes we would have fighter escorts. If I was shot at before I got near the target, it resulted in more evasive action.

Pat: Tell me about the mission.

George: The mission was assigned normally two days, but at least one day, in advance. The number of airplanes on one mission could be forty to fifty, eighty, ninety.   We could have one or two or three missions a day. The average round trip was from Italy to Italy (US forces were mostly in Italy) or from Italy to Germany. On a mission, the Germans tried not to shoot the first planes in front of the box. The middle ones made a better target. Anti-aircraft shells did most of the damage. The 340th became so accurate they were given bridges to hit, since those were the hardest. They earned the nickname of the "Bridge Busters." Initially after fifteen to twenty-five missions were flown, men were sent home.

Pat: Type of bomb load?

George: There were three or four different types of bombs. We were given whatever plane was ready, told the type of bomb load and what the target was. There was not much airpower fight from Italians.   From them most damage was from their ground troops. We avoided bombing the Abbey at Montecassino, for obvious reasons, until it was confirmed that Germans were using the Abbey to relay info. I was the lead pilot to hit the Abbey. The B-25 could fly with only one engine. It was not so stable, but it was OK. Its ability was reduced.

Pat: Describe your return to base.

George: We returned to base and landed in the same flight formation of 3, 3. Same position as when we left. When we got near the airstrip the aircraft spread, one behind the other, about twice the plane's length. The airstrip can handle three planes abreast but flared out.[3]

The conversation began to drift around other subjects as friends meandered over to share a drink, a photo or a story. To share, most accurately, a closeness. George, with his Shirley, in her turquoise silk dress, beside him, absorbed the evening and was content.

George and Pat

PART II

MEET THE HEROES

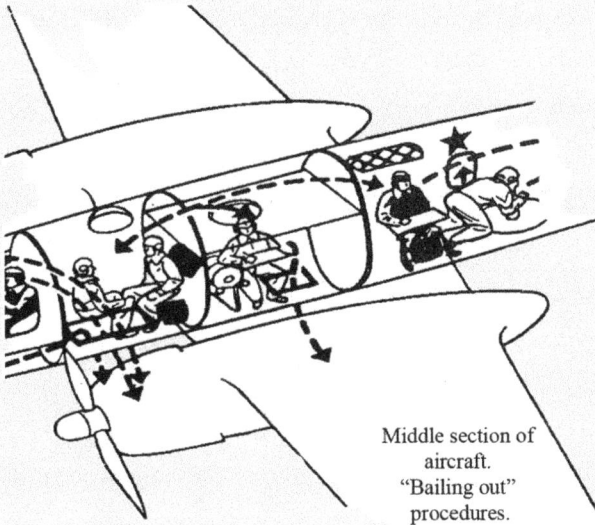

Middle section of
aircraft.
"Bailing out"
procedures.

WATCH HOW LIFE SEEDS ART

It is from the truths, the facts, and the solid ground of life that *Catch-22* sprouted. And while, under Heller's masterful touch, great fiction blossomed, still this fact/fiction dependent relationship is undeniable. The following pages show how these forces play with and against each other.

The characters of *Catch-22* are strong and unique and independent, but many of them owe the seeds of their very existence to the also strong and unique and independent men of the 340th Bomb Group. Now, introductions are necessary...

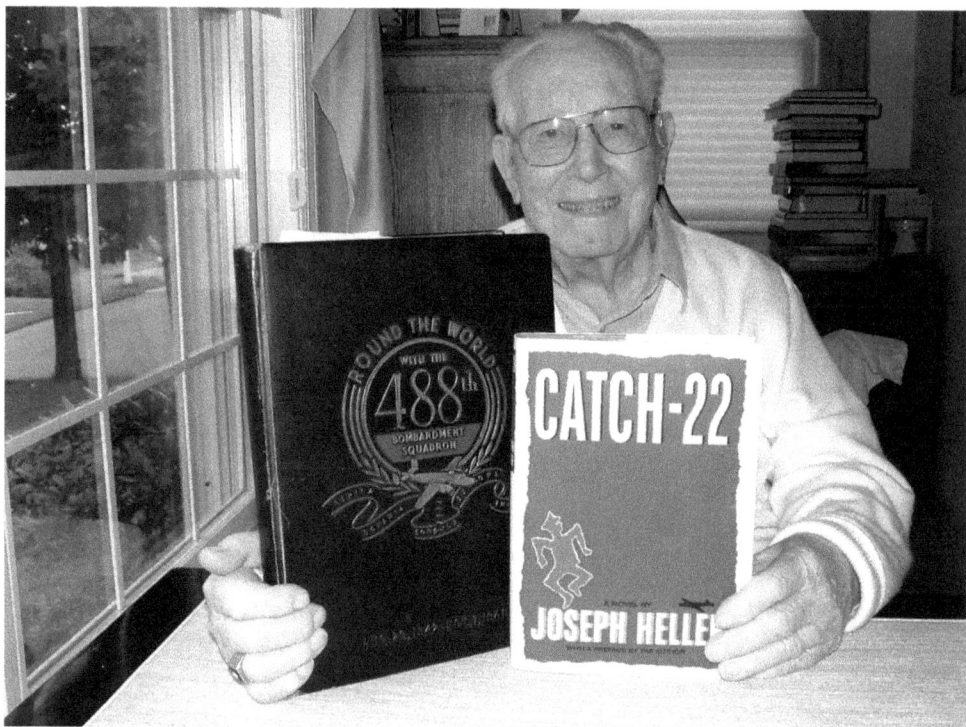

George with his 488th Squadron Yearbook and *Catch-22.*

BOOK CHARACTERS & MEN OF THE 340TH

General Dreedle	General Robert D. Knapp
Colonel Cathcart	Colonel Willis F. Chapman
Chaplain Tappman	Chaplain James H. Cooper
Doc Daneeka	Captain Benjamin J. Marino, M.D.
John Yossarian	Lt. Joseph Heller
Chief White Halfoat and Havermyer	Captain Vincent Myers
Hungry Joe	Joseph Chrenko, pilot
Major Major Major Major	Major Major and Captain Randall C. Cassada
Milo Minderbinder	Benjamin Kanowsky, pilot, mess officer, etc
Captains Piltchard and Wren	Capt. Fred W. Dyer and Capt. George L. Wells
Soldier in White	Those killed in the war
Major __de Coverley	Major Charles J. Cover
Douglas Orr	Douglas Orr, Bomb./Nav. and Edward Ritter, pilot
Luciana	Girl Heller met in Rome
Kid Sampson	Bill Simpson, pilot
Major Danby	Major Joseph Ruebel
Col. Moodus	Lt. Robert Knapp, Jr.

...when Yossarian climbed down
few steps of his plane naked, in a s
of utter shock, with Snowden sme
abundantly all over his bare heels
toes, knees, arms and fingers,
pointed inside wordlessly. .(

"I don't want to wear a uniform any m

Milo ... found him sitting up a tr
small distance in back of the qua
little military cemetery at wh
Snowden was being buried." (

*"Yossarian was a lead bombardier who had been demoted because he no longer
gave a damn whether he missed or not. ."*(p.29)

YOSSARIAN

LOOSELY   BASED UPON ➡

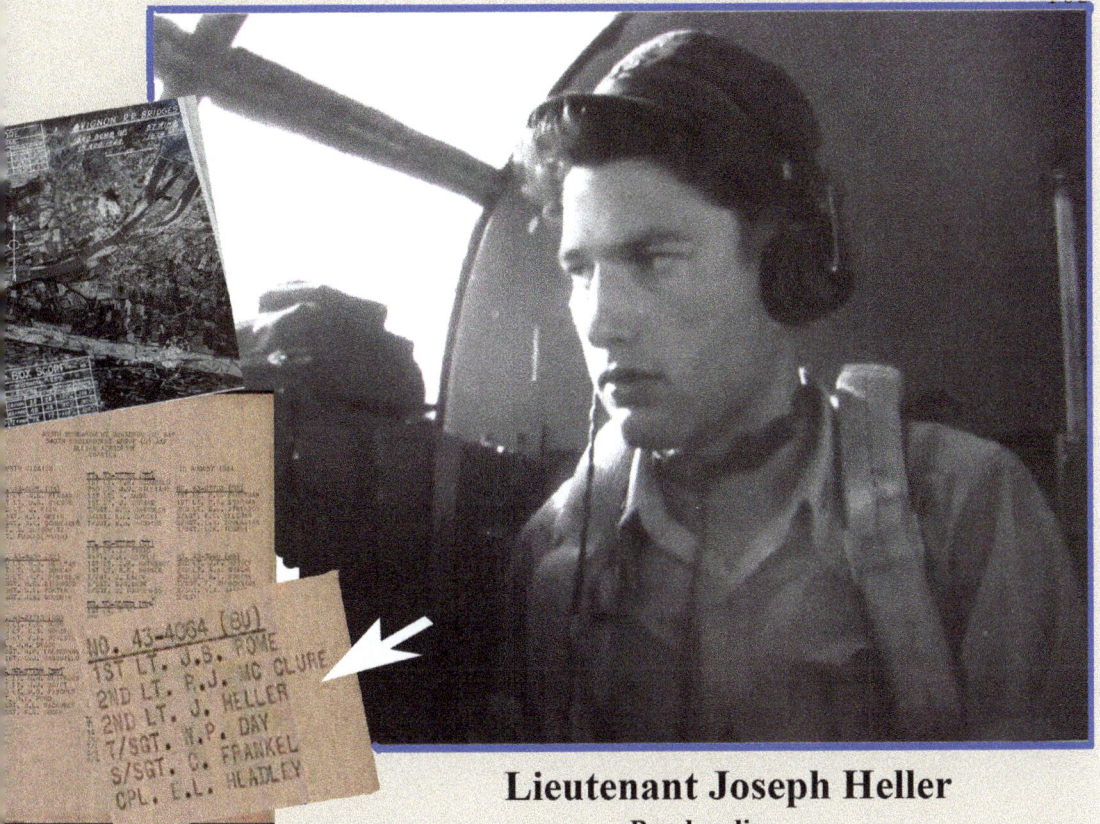

THIS PAGE DECLASSIFIED IAW EO12958

Lieutenant Joseph Heller
Bombardier
340th Bomb Group

Decorations:
Air Medal
Presidential Unit Citation

War Diary August 15, 1944 Mission to Abignon Railroad Bridges (Mission #289)

The
True
Story
- War
Diaries

488th. From Mission report
 FLAK: Heavy, intense and accurate coming from the aerodrome south of town - Red hanging puffs. New guns at vS-9380, to 9980. Northern approach of both bridges in target area. 486th plane 3,000 ft. down.

OBSERVATIONS; At Y-7875 8D was seen to crash at 1700 hrs. 3 chutes at &7;075. 5 chutes seen coming from plane. 1 failed to open properly. Left engine was on fire and right engine was out. On bomber 8U on August 15, 1944 Lt. Joseph Heller poured sulfa into the open wound on S/Sgt. Carl Frankel's leg, bandaged it, and stayed with him until the plane landed at Alesan Air Base where Sgt. Frankel would have been transported to the field hospital at Cervione not far from Alesan.
Frankel survived the injury.

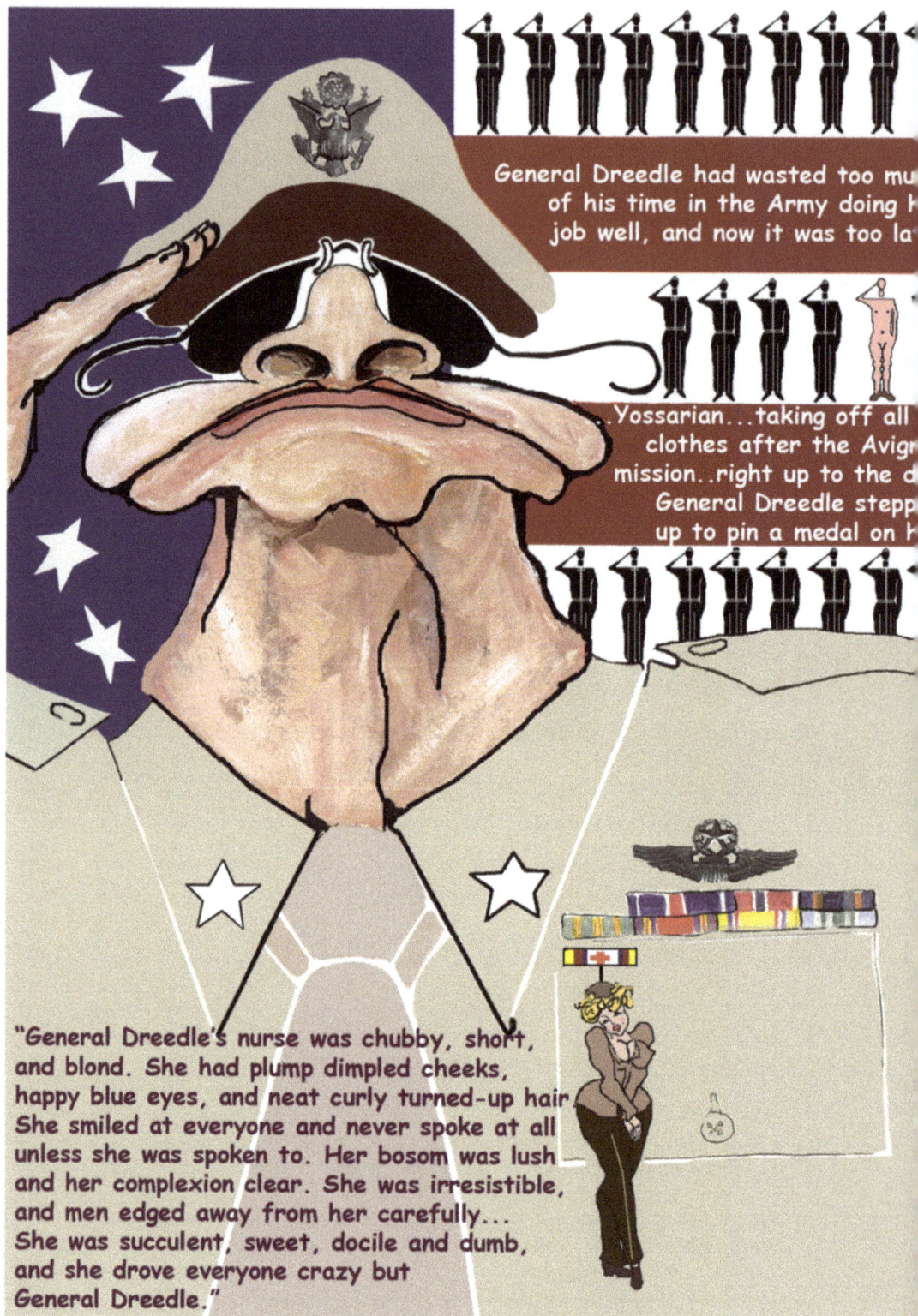

General Dreedle had wasted too mu
of his time in the Army doing h
job well, and now it was too la

..Yossarian...taking off all
clothes after the Avign
mission..right up to the d
General Dreedle stepp
up to pin a medal on h

"General Dreedle's nurse was chubby, short,
and blond. She had plump dimpled cheeks,
happy blue eyes, and neat curly turned-up hair.
She smiled at everyone and never spoke at all
unless she was spoken to. Her bosom was lush
and her complexion clear. She was irresistible,
and men edged away from her carefully...
She was succulent, sweet, docile and dumb,
and she drove everyone crazy but
General Dreedle."

GENERAL DREEDLE

LOOSELY   BASED UPON ➡

General Robert D. Knapp

Commanding General of the 57th Bomb Wing

Chapman, left, Knapp, 2nd from right, Bailey, right, relaxing over a game of bridge...

Decorations

Distinguished Service Medal
Silver Star
Distinguished Flying Cross
Bronze Star
Air Medal with seven Oak Leaf Clusters
Polish Cross of Valor
Commander Order of the British Empire
Greek Cross of Valor
French Croix de Guerre

The True Story
- Robert D. Knapp

"One warm evening I decided to take a spin- we were encouraged to fly as often as possible. Well, I took off, carelessly forgetting to check the gasoline tank. Airborne, I tried a few figure eights and rolls when suddenly I spotted an airplane in the river below me with its tail sticking up above the water. I circled around looking for the pilot and then I spotted him with his head on the side of the cockpit, obviously in pain.

I was about to head back to the base to report the accident when suddenly I ran out of gas! I went into a stall and spun into the river right long side the other plane. Fortunately I was not injured seriously. I suffered a broken nose and few facial and body cuts, but no major damages. I managed to get out of my plane and wade over to the other pilot, his name was Kennedy, a man who shared my hanger. I pulled him out of his plane and carried him to a nearby road where I hailed a passing car. Our Good Samaritan drove us to the city hospital.

In a few weeks (1917), we completed our courses and were commissioned as 2nd Lts. Getting those wings had to be one of my proudest moments."

"Fly Low and Slow, Robert", The Men of the 57th, January 1979

"Col. Cathcart had courage and never hesitated to volunteer his men for any target available. No target was too dangerous for his group to attack."

"Col. Cathcart was conceite because he was a full colon with a combat command at t age of only 36; and Col.Cathca was dejected because, althoug he was already 36, he was still only a full colonel."

COLONEL CATHCART
LOOSELY   BASED UPON ➡

ng after returning
mission.

n reads:
erations
mbardment Group
Medium
m Group There Is
ct of the USA"

seat of B-25
igarette holder

Receiving Distinguished Flying Cross
from General Knapp

Decorations

Distinguished Flying Cross
Bronze Star Medal
Air Medal with four Oak Leaf Clusters
Joint Services Commendation Medal
Distinguished Unit Citation
American Defense Service Medal
American Campaign Medal
European-African-Middle Eastern
Campaign Medal with two Silver Stars
WWII Victory Medal
National Defense Services Medal
Air Force Longevity Service Award
with one Silver Star and one Bronze
Oak Leaf Cluster
French Croix de Guerre with Palm

Colonel Willis F. Chapman
Commander, 340th Bomb Group
1944-1945

THE TRUE STORY

Bill Chapman

"I assumed command of the 340th Bomb Group, flying B-25 Mitchell Bombers on 16 March 1944. Five days later I faced quite a shock. The best first hand description is in the following V-Mail letter to my wife.

March 20, 1944

Sweetheart- Still no letters. I don't know why the mails have to get all tied up at this late stage. It should be well organized. There is a little volcano over here which is putting on a grand show. I've never seen anything like it. I can watch it from the window of my trailer. It is magnificent. You can see the individual explosions which are continuous and watch the pieces go up and then fall back down the side of the cove. The large lava flow is on the other side toward the sea and has caused many homes to be evacuated. The explosions can be heard all the time here and there is a continuous rain of fine particles coming down all the time which sounds like rain. I only wish I had my movie camera here with some Kodachrome film. The activity has been increasing for the last three days and tonight it is really going to work. I like my new job a lot. I wouldn't swap it for anything else in the Air Force now. These B-25s are really wonderful airplanes. I'm anxious to know if you've moved yet, Hon. The suspense is terrific. I hope you are getting my letters all right. Give the girls a big kiss for Daddy and tell them that he is really trying to get this war over with in a hurry now.
All my love,
Darling -Bill

he shock came about 5:30 PM on 21 March when Mt. Vesuvius erupted violently 15-20,000 feet high. The wind from the northwest blew hot ash and debris over our airdrome (Pompeii) covering everything to a depth of 18-20 inches. It crazed the plexiglas in the aircraft canopies, ruined the canvas control surface and dented some of the metal surfaces, making all eighty-seven B-25s unflyable. The men were not certain if they would be mummified by morning or not. We started evacuation to Guendo, near Salerno, on the 22nd."

"When the Germans left the city, I rushed out to welcome the Americans with a bottle of excellent brandy and a basket of flowers.

The brandy was for myself, of course, and the flowers were to sprinkle upon our liberators.. -a sordid, diabolical old Italian man.

Major __ de Coverley was a splendid, awe-inspiring, grave old man with a massive leonine head and an angry shock of wild white hair that raged like a blizzard around his stern, patriarchal face.

MAJOR ____ DE COVERLEY

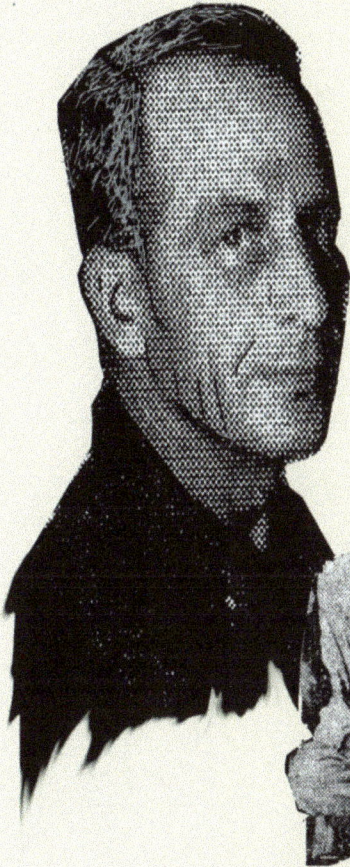

Major Charles J. Cover
Squadron Executive Officer
340th Bomb Group

"Jerre Cover was an absolute gentleman through and through. He was older than most of the other men, fifty-ish. He had grey hair, even at that time, with a face that was tan and deeply lined. He was more of a father figure than a military figure to the men. He was lean, alert and a tough cookie."

- Forrest Wells and Bill Chapman

SECRET

488TH BOMBARDMENT SQUADRON (M) AAF
340TH BOMBARDMENT GROUP (M) AAF
ALESAN AERODROME
CORSICA

FIRST MISSION   24 MAY 1944

NO. 43-4013 (8F)
CAPT. G.L. WELLS
CAPT. H.B. HOWARD
2ND LT. C.W. HATHAWAY
2ND LT. P. KOK
T/SGT. F.J. LAPOINT
S/SGT. C.A. WOOD
S/SGT. R.K. SHOWALTER

NO. 43-27522 (8N)
2ND LT. S. ASWAD
2ND LT. M.A. LINDHOLM
2ND LT. F.S. NEWHARD
S/SGT. J. LAZOR
S/SGT. C.J. MASCUILLO
SGT. W. WOYTEK

NO. 43-27474 (8R)
2ND LT. P.K. WEATHER
2ND LT. H.C. GROSS
1ST LT. E.A. LINK
S/SGT. O.O. CHANDI
SGT. W.E. PORTER
SGT. A.E. ROSIN

NO. 43-27700
2ND LT. H.C.
2ND LT. E.C.
2ND LT. R.R. BURG
T/SGT. B. GREENBAUM
S/SGT. K.L. ADAMSON
T/SGT. W.H. HUDGINS

NO. 43-27857 (8P)
1ST LT. G.V. SMITH
2ND LT. J.F. MUMMEY
2ND LT. C.O. SMYRE
S/SGT. R.J. MARTIN
S/SGT. T.H. HEATH
S/SGT. J.D. REYNOLDS

NO. 43-27532 (8A)
2ND LT. J.G. CHRENKO
2ND LT. O.R. WILSON
2ND LT. J. HELLER
S/SGT. J. GOODELL
SGT. A. KRAUSE
SGT. C.G. BEATTY

NO. 43-27713 (8D)
2ND LT. B.C. DIEKMAN
2ND LT. E.G. SALLEN
2ND LT. W.A. DAVIDSON
2ND LT. E.F. MURRAY
T/SGT. A.J. ALLEN
S/SGT. M.A. KING
S/SGT. R.M. SCOTT

NO. 43-27702 (8M)
2ND LT. A.D. HEMSTAD
F/O A.N. RYCK
2ND LT. J. NORMAN
S/SGT. A. VANDERMUELEN
SGT. H.E. BARTELL
CPL. E.D. ARNOLD

NO. 43-27729 (8S)
2ND LT. W.B. REAGAN
2ND LT. J.B. ROME
2ND LT. R.H. PINKARD
S/SGT. A.L. GREEN
SGT. N.E. KLINKNER
SGT. A.J. BERTAGNA

NO. 43-27769 (8T)
2ND LT. C.R. HAGERMAN
2ND LT. E.W. MCDONALD
2ND LT. M.V. MISEVIC
T/SGT. C.H. MCARTHUR
SGT. C. KETTERINGHAM
S/SGT. J.E. SMITH

NO. 43-27754 (8)
2ND LT. P.R. FERRYMAN
1ST LT. G.W. CLIFFORD
2ND LT. A.R. PANKO
S/SGT. J.E. BRONSON
S/SGT. R. BLAND
S/SGT. J.L. LYONS

NO. 43-4064 (8-)
2ND LT. H.C. MCELROY
2ND LT. M.G. DUNCAN
F/O D.L. ATKINSON
S/SGT. A. SCHROEDER
SGT. H.L. RACKMYER
SGT. J.L. MARKEY

FLEW WITH 486TH
NO. "990"
2ND LT. T.E. JONES
2ND LT. J.E. COOPER
2ND LT. W.O. FISCHER
S/SGT. W.P. DAY
SGT. B.E. GORSKI
SGT. M.B. WNN

John G Murphey

JOHN G. MURPHEY,
1ST LT., AIR CORPS,
ASS'T. S-2, 488TH BOMB SQ.

SECRET

THIS PAGE DECLASSIFIED IAW EO12958

This mission included:
George Wells   (Captain Wren)
Joe Chrenko   (Hungry Joe)
Joe Heller   (Yossarian)

Dyer, center,
with friends

Major Fred W. Dyer
Assistant Group Operations Officer
340th Bomb Group

Decorations
Distinguished Flying Cross
Purple Heart
Distinguished Service Cross
Silver Star
Air Medal with 15 clusters

THE TRUE STORY
- Fred Dyer

"My most memorable mission? I could hardy forget the one in July 17, 1943, over the Herman Goering Division at Randazzo, Sicily, when I got shot down. On this mission we got hit by ack-ack approaching the target at thirteen thousand feet. I was number three in the box. We had a big hole in the right wing as if an 88 mm shell had gone right through it, the nose was blown off, the right engine was running away and all the engine controls were ineffective. We kept on going over the target and managed to drop our bomb load at six thousand feet. We could see the rest of the formation heading for home. The airplane wasn't flying very good by this time and we kept losing altitude. At twenty-five hundred feet it was time to bail out. John Falwell, my copilot, Pete Cusintine, the engineer and I got out. One crewmember went down with the plane. John, Pete and I landed in a burned over wheat field and managed to evade capture. One other crewmember was captured. We had apparently landed in the midst of a tank battle between the German troops and the British 51st Highland Division. Later on that afternoon we were picked up by a tank of the 51st Division and taken to their field headquarters where we spent the night listening to shells whistle over. We talked to one British major who had been in the Middle East for three years at that time. What a battlehardened, tough outfit this was. After a week's odyssey we finally got back to Hergla, Tunisia. That was my twelfth mission."

"Captain Piltchard and Captain Wren, the inoffensive joint squadron operations officers enjoyed flying combat missions and begged nothing more of life and Col. Cathcart than the opportunity to continue flying them. They had flown hundreds of combat missions and wanted to fly hundreds more. They assigned themselves to every one. Nothing so wonderful as war had ever happened to them before; and they were afraid it might never happen to them again."

CAPTAIN PILTCHARD & CAPTAIN WREN

 LOOSELY   BASED UPON

They were shifty, subservient men who were comfortable only with each other and never met anyone else's eye, not even Yossarian's eye at the open-air meeting they called to reprimand him publicly for making Kid Sampson turn back from the mission to Bologne.

Well, we finally got to Bologna today, and we found out it's a milk run. We were all a little nervous, I guess, and didn't do too much damage. Well, listen to this. Col. Cathcart got permission for us to go back. And tomorrow we're really going to paste those ammunition dumps. Now, what do you think about that?"

to prove to Yossarian that they bore him no animosity, they even assigned him to fly lead bombardier ... (p.144)

In a 1942 photograph, George Wells sits in the front seat of a Stearman biplane during flying school.

In front of B-25, 1944
Wells, waiting to fly a secret mission
for the 12th Air Force into Italy to
drop off agents, pictures, and supplies
This flight was aborted at the last
minute.

Fred Dyer, center and Gee

Major George L. Wells

Assistant Group Operations Officer

340th Bomb Group

THE TRUE STORY
-George Wells

"Having been a Field Artillery officer prior to becoming a pilot in the Air Corps, I continued to wear my garrison cap as I did in the Field Artillery. I wouldn't wear the earphones (head set) over the outside of the cap which would crush the crown. A crushed garrison cap in the Air Corps was the 'in thing', the mark of a pilot. Well, I had a brand new garrison cap when I left for overseas with a real good military appearance. It had a great Field Artillery look but not an Air Corps look. When I joined the 488th Squadron I didn't wear it much, but when I did, I received a lot of kidding. In late April, 1944 I got a chance to fly down to Alexandria and Cairo for a week's rest with two other officers and five enlisted men. Homer Howard and I rotated on each leg of the flight from pilot to copilot position. On the second leg I was in the pilot's seat with my headset on and my great garrison cap sitting on the top of my head over the headset. By this time we were over the desert and it was very hot in the B-25 (no air-conditioning). We had opened the two sliding side windows in the cockpit to suck some of the heat out. I had been flying the plane for 11-12 hours and I turned it over to Homer in the copilot's seat. As I leaned to the left my cap just flew straight out the small opening. Some Arab must have found a beautiful cap and the guys never let me forget the consequences of my refusal to accept 'The Crushed Hat.' "

Decorations

USAF Outstanding Unit Award
Presidential Unit Citation
National Defense Service Med
with one Bronze Star
Air Force Longevity Service A
with one Silver Oak Leaf Clus
World War II Victory Medal
European-African-Middle Eas
Campaign Medal with seven
Battle Stars
American Campaign Medal
American Defense Service Me
Joint Service Medal
Air Force Commendation Med
Joint Service Commendation M
Air Medal with twelve Oak Lea
Clusters
Distinguished Flying Cross wi
one Oak Leaf Cluster
The Silver Star

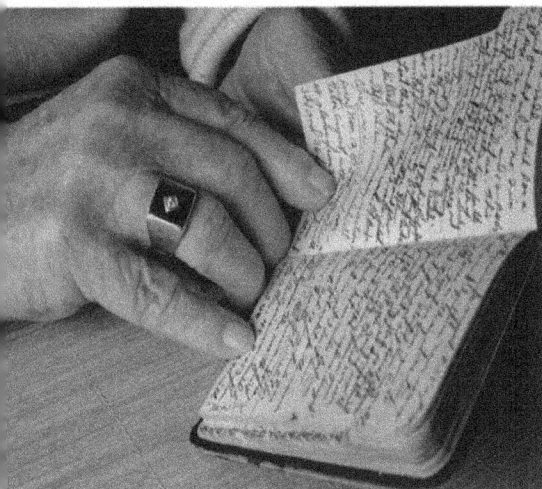

At age 90, George reminisces as he thumbs through the pages of his well-worn mission book.

Upon his return to home base from a mission, George would pull out this small black book and routinely log each of his 102 missions. The entries, brief and usually devoid of emotion, fleshed out the true story of a pilot's circumstances. They would show, curiously, how sometimes the more difficult missions were the least storied.

For example:
"Mission 95-December 22, 1944
Formation commander with 489th on target at Lavis on the Brenner Pass line. Ran into lots of ack-ack and picked up quite a few holes. Really was a cold ride for four and a half hours."

The chaplain felt his face flush.

"I'm sorry, sir. I just assumed you would want the enlisted men to be present (for prayers) since they will be going on the same mission."

"Well, I don't.

"They've got a God and chaplain of their own, haven't they?"

"No, sir."

"What are you talking about? You mean they pray to the same God we do?"

"Yes, sir."

"And He listens?" (P.191) - with Col. Cathcart

He toyed unfamiliarly with the tiny corncob pipe the affected selfconsciously occasionally stuffed with tobacco and smoked. (P.2...

CHAPLAIN A.T. TAPPMAN

LOOSELY   BASED UPON ➡

Capt. Cooper & Col. Chapman
488th Club, Corsica

Foggia, Italy, 1943.
Note pipe.

James H. Cooper
Chaplain
340th Bomb Group

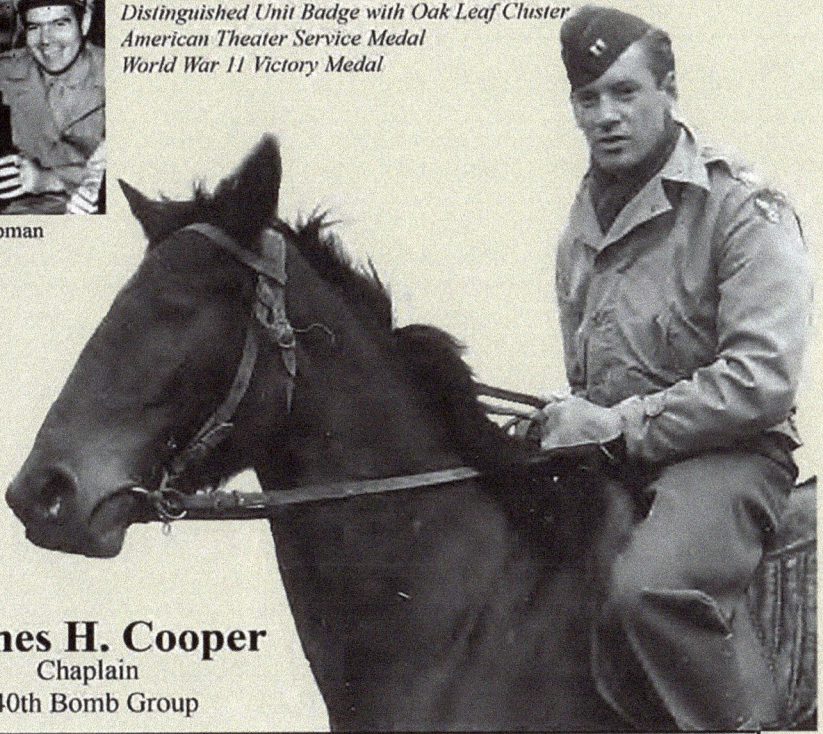

The True Story
- Jim Cooper

"On August 24, 1979, exactly 1900 years had passed since Vesuvius blew up with what is now known as the most destructive eruption in recorded history. The city of Pompeii, Italy near the bay of Naples disappeared completely. It is almost impossible for us to imagine the horror and panic of such a catastrophe. Yet there are many Air Force service men from World War II who have a very good idea of the lethal power of Vesuvius!

It all began quietly enough in March 1944. Our bomb group, the 340th, and other Air Force groups, were stationed around the base of Vesuvius, engaging in bombing Italy. Our airfield, near Pompeii, was bulldozed out of the lava and ash deposited 19 centuries before.

Pilots returning from missions day or night could easily find our airstrip by locating Vesuvius! In daylight the white wisps of smoke rising from its cone, and a red glow at night from the crater, made an easy landmark.

Two other officers and I drove a jeep up the mountain as far as the road went. We then walked to the top. The terrain was rough and quite ugly. We were amazed at the raw, jagged, and awesome appearance of the volcano's cone. From fissures, a slow bubbling red flow of lava, while not threatening, persisted slowly toward the outer rim.

A few enterprising native children were dipping out small globs of lava on sticks, pressing small Italian coins into the soft but quickly hardening liquid stone, and charging a dollar.

Two days later there appeared to be more smoke than usual coming out of Vesuvius, and at night there was an obvious red glow at the top that had not been evident before. The next morning we returned to the top. This time we had to pick our way around and over swollen streams of molten stone. You could walk on the spongy, black surface of the fast cooling lava but underneath was a deep red glow. As these streams struck trees or bushes, there was a match-like spurt of flame, then the tree, or twig simply disappeared with a little puff of smoke.

We were still not alarmed, for the slowly advancing streams seemed to pose no serious problem for the farms and villages further down the mountain .

(Shortly) the falling debris...ashes, cinders of great size, and acrid smoking clinker, made the wearing of helmets mandatory. Natives living on the higher elevations of the mountain in villages and farms streamed down the volcano side taking refuge in churches, where there was much wailing and praying. There were some small villages, farms and vineyards destroyed. Vesuvius was erupting!

In the dark before dawn we couldn't assess our damage, but it became quite clear when morning arrived. Every airplane was riddled with gapping jagged holes in wings and fuselages. Ashes were built up to the top of the landing gear. For those sleeping in tents, it had been a frightful night. In their tattered and sieve-like condition, tents were no protection.

Our planes were thus ruined, and with a volcano of indeterminate length raging above us, a quick decision to evacuate was ordered. As quickly as possible we fled as the Pompeians had done 19 centuries before. Only in our case we fled in trucks and jeeps, going down the coast for many miles to an area that had once been a Greek colony, and where still stood a Greek temple... Paestum.

The irony of it all was that the Axis Powers had been trying to put us out of business for a long time. But what they had not been able to do in many months, Vesuvius accomplished in one night."

WATCH
Bill never chose to wear his "good" watch, a graduation gift, on a mission. He always wore this "mission watch" in case he got shot down.

W.F. CHAPMAN

LEATHER BOMBER JACKET
Still resting in the pocket is his name tag and this box of NODOZ.

Essentials belonging to Bill Chapman

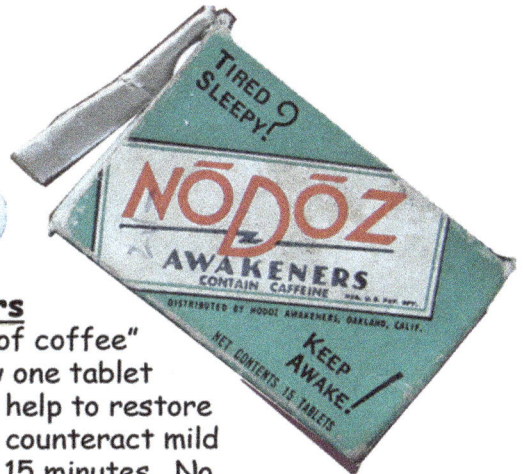

TIRED SLEEPY!

NŌDŌZ

AWAKENERS
CONTAIN CAFFEINE

DISTRIBUTED BY NODOZ AWAKENERS, OAKLAND, CALIF.

NET CONTENTS 15 TABLETS

KEEP AWAKE!

NoDoz Awakeners
"Harmless as a cup of coffee"
For adults: Swallow one tablet with water. Should help to restore mental alertness or counteract mild alcoholic effects in 15 minutes. No compensating depression. Four tablets within a period of twelve hours should be sufficient.

"... LEG BONE'S CONNECTED TO DE BODY BONE..."

Doc Daneeka was a very neat, clean man whose idea of a good time was to sulk. He had a dark complexion and a small, wise, saturnine face with mournful pouches under both eyes. He brooded over his health continually and went almost daily to the medical tent to have his temperature taken by one of the two enlisted men there who ran things for him practically on their own. They could never find anything wrong with him. He was thinking of having them both transferred back to the motor pool and replaced by someone who could find something wrong. (p.32)

DOC DANEEKA

LOOSELY   BASED UPON ➤

In his identifiable moccasins

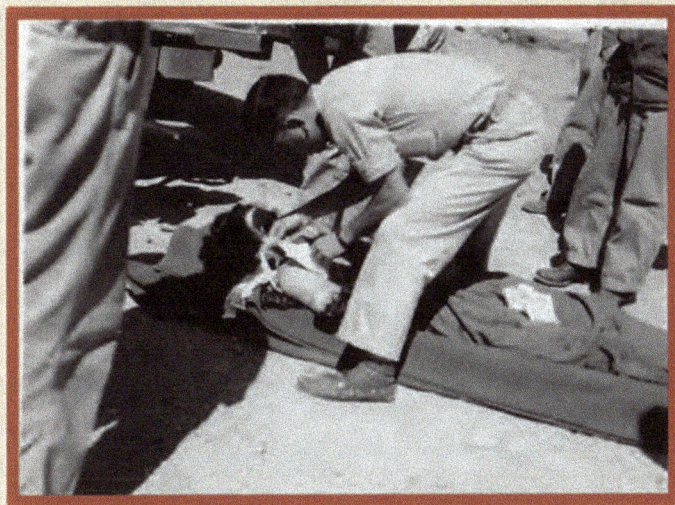

Doc

Captain Benjamin J. Marino, M.D.

Flight Surgeon
488th Bomb Squadron
340th Bomb Group

The True Story
- Robert B. Marino, son

"I made contact with Mr. Heller through his literary agent. I asked him if my father was the flight surgeon mentioned in his book. He stated he was. I also related a story my father told me. Whenever a new group of flight crews returned from their first missions, there would be a line out in front of the medical tent. 'Doc, it's dangerous up there, you've got to get me out of here." In fact, my father stated that several members threatened him with bodily harm if he didn't act favorably on their transfer requests. This was truly Catch-22."

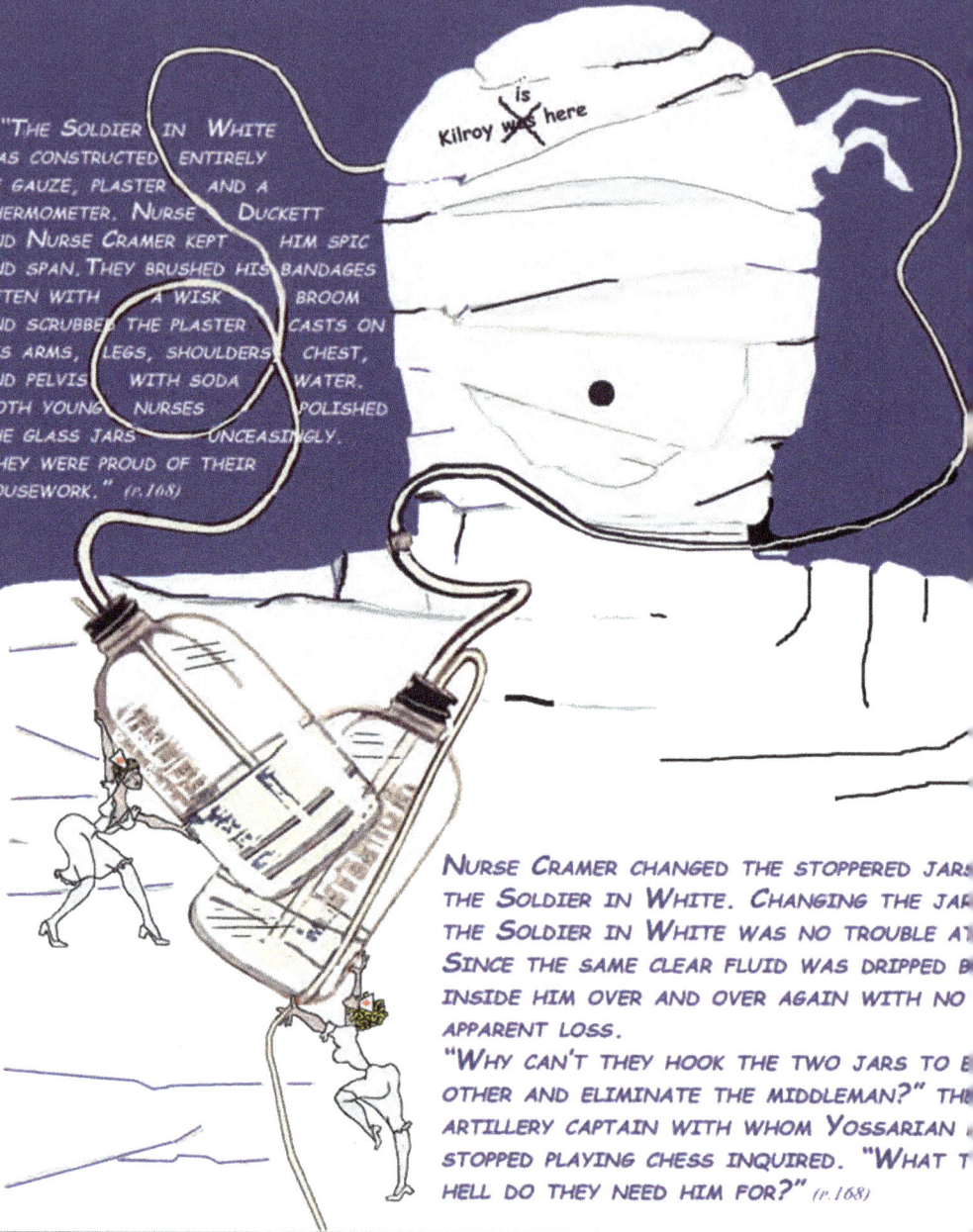

Kilroy ~~was~~ is here

"THE SOLDIER IN WHITE WAS CONSTRUCTED ENTIRELY OF GAUZE, PLASTER AND A THERMOMETER. NURSE DUCKETT AND NURSE CRAMER KEPT HIM SPIC AND SPAN. THEY BRUSHED HIS BANDAGES OFTEN WITH A WISK BROOM AND SCRUBBED THE PLASTER CASTS ON HIS ARMS, LEGS, SHOULDERS, CHEST, AND PELVIS WITH SODA WATER. BOTH YOUNG NURSES POLISHED THE GLASS JARS UNCEASINGLY. THEY WERE PROUD OF THEIR HOUSEWORK." (P.168)

NURSE CRAMER CHANGED THE STOPPERED JARS THE SOLDIER IN WHITE. CHANGING THE JAR THE SOLDIER IN WHITE WAS NO TROUBLE AT SINCE THE SAME CLEAR FLUID WAS DRIPPED B INSIDE HIM OVER AND OVER AGAIN WITH NO APPARENT LOSS.

"WHY CAN'T THEY HOOK THE TWO JARS TO E OTHER AND ELIMINATE THE MIDDLEMAN?" TH ARTILLERY CAPTAIN WITH WHOM YOSSARIAN STOPPED PLAYING CHESS INQUIRED. "WHAT T HELL DO THEY NEED HIM FOR?" (P.168)

THE SOLDIER IN WHITE
Tragically influenced by ➤

MOST, BUT NOT ALL, OF THE MEN UPON WHOM JOSEPH HELLER BASED HIS FICTIONAL CHARACTERS IN *CATCH-22* CAME FROM THIS **488TH** SQUADRON, INCLUDING HELLER HIMSELF.

NICHOLAS ALBANESE   ARTHUR J. BERTAGNA   CHARLES S. BORDEN   SAMUEL G. BRADLEY   ROBERT J. BROWN   WILLIAM J. BROWN   JAMES C. BURKHUS, JR.   BERNARD L. CORBIN

ROBERT E. DEAN   JAMES C. DE LUCCA   AUSTIN E. DUNAWAY   GEORGE C. GARSKE   FREDERICK, C. GREENIG   GEORGE GREGORY   LEE D. HANLAN   CLARENCE E. HAYNES

ROBERT J. JACKSON   ROBERT KEEDY   ARNOLD L. KRAUSE   JOHN M. LEVERETTE   HARRY F. LUEDERMAN   HARRY I. LUTH   JOSEPH A. McGINNIS   RICHARD MILLER

488TH
SQUADRON COMBAT LOSSES
6 APRIL, 1943 - 26 APRIL, 1945

REMEMBRANCE

KILLED IN ACTION	 47
MISSING IN ACTION	 12
ACCIDENTAL DEATHS	 4
DIED	 1

ALBERT H. MITCHELL   MATTHEW C. MORRISON   HARDY D. NARRON   ROBERT F. NOBLE   WILLIAM B. PELTON   ROY H. PINKARD   AUBREY B. PORTER

GEORGE C. PRICE   HOWARD G. REICHARD   JAMES D. REYNOLDS   JAMES C. RICE   JOHN E. RILEY   ROBERT G. SCHLITTLER   LOUIS H. SCHMIDT

DOYLE G. SHIPLEY   WILLIAM Y. SIMPSON   MELARD H. TAFOYA   GRANT R. THORSTED   ARTHUR VANDERMEULEN   ERWIN H. WILKE   JOHN P. WILKINSON

HARRY S. WILSON   FREDERICK J. WOHLSTEIN   PAUL M. ZIC   WALTER F. ZIEGLER

HAROLD H. HAMMOND	BYRON D. KING
ROBERT E. GARDNER	JAMES A. BURGER
DAVID R. POWERS	FRED B. HICKS
ROY E. ROGERS	ARTHUR L. ALLEN
JAMES E. COOPER	LAWRENCE W. KAHL
ALVIN H. YELLON	BLAINE G. THOMPSON
EARL C. MOON	JAMES H. CARDWELL
JAMES H. DYSON	GEORGE W. HAMMOND, JR

OF PILOT "KID" SAMPSON

REICHARD, GARSKE, AND DEAN, SURFACING IN THE TRUE STORY OF CATCH-22, WERE SHOT DOWN IN THE SAME B-25.

PAIRED WITH "KID" SAMP WHEN HE, AS TAIL GUNNER, WAS CUT OUT OF HIS PLANE

The Dead Man in Yossarian's Tent

"Almost without realizing it, Sergeant Towser had fallen into the habit of thinking of the dead man in Yossarian's tent in Yossarian's own terms - as a dead man in Yossarian's tent. In reality, he was no such thing. He was simply a replacement pilot who had been killed in combat before he had officially reported for duty. He had stopped at the operations tent to inquire the way to the orderly-room tent and had been sent right into action because so many men had completed the thirty-five missions required then that Captain Piltchard and Captain Wren were finding it difficult to assemble the number of crews specified by Group. Because he had never officially gotten into the squadron, he could never officially be gotten out and Sergeant Towser sensed that the multiplying communications relating to the poor man would continue reverberating forever.

His name was Mudd. To Sergeant Towser, who deplored violence and waste with equal aversion, it seemed like such an abhorrent extravagance to fly Mudd all the way across the ocean just to have him blown into bits over Orvieto less than two hours after he arrived. No one could recall who he was or what he had looked like, least of all Captain Piltchard and Captain Wren, who remembered only that a new officer had shown up at the operations tent just in time to be killed and who colored easily every time the matter of the dead man in Yossarian's tent was mentioned. The only ones who might have seen Mudd, the men in the same plane, had all been blown to bits with him."

"That story of the dead man in the tent, it's sort of factual. He was shot down before his orders came in. He just lay around until the red tape caught up. Then we post-dated his orders, and got him out of there.

The human mind is funny. You reach a point where it shuts things out. We thought it was funny, where's your buddy, and all there was was an empty bunk. People were put in a position of doing things they didn't want to do."

- Ben Kanowsky (Milo Minderbinder)

ON MARCH 22, 1944, THE 340TH BOMB GROUP, 57TH BOMB WING, LOST 87
B-25 MITCHELL BOMBERS UNDER THE ERUPTION OF MT. VESUVIUS IN ITALY.

IN 1984, JOSEPH N. KLEIN, JR. COMPLETED THIS PAINTING COMMEMORATING
THE 40TH ANNIVERSARY OF THIS EVENT AS A GIFT TO HIS FATHER,
JOSEPH N. KLEIN, SR. JOE, SR. FLEW SEVERAL OF HIS MISSIONS IN THIS
PARTICULAR AIRCRAFT.

Oo ioooooooooooooooooooooooh

hooooo hooooo

Ooooh oooh ooooh

Ooooooh

"Sick with lust Yossarian moaned Oooooooooooooooooo...
"I run a fighting outfit and there will be no more moaning in
this group as long as I'm in command." (Dreedle)
It was clear to everybody but major Danby, who was still
concentrating on his wrist watch and counting down the seconds
aloud, "... four ... three ... two ...one ... time!" called out Major
Danby and raised his eyes triumphantly to discover that no one
had been listening to him and that he would have to begin all
over again.
"Oooo," he moaned in frustration.
General Dreedle whirled around in a murderous rage.
"Take him out and shoot him!" (p.217)

MAJOR DANBY

LOOSELY   BASED UPON ➡

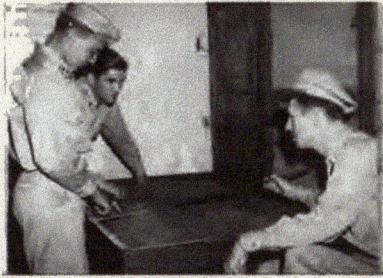
At the War Desk

B-25 crew. Joe, back row, second from right.

Receiving medal from General Robert Knapp, who is flanked by a pleased Col. Chapman.

Major Joseph W. Ruebel
488th Adjutant
340th Bomb Group

The True Story
- Joe Ruebel

"A seldom mentioned occupational hazard encountered in the air war over Italy was the outside temperature at B-25 bombing altitude in the winter time. The gauge in the aircraft often indicated an outside air temperature reading of -50 degrees. As this reading was the bottom of the scale on the gauge, the actual temperature might have been lower.

Unfortunately, any provision for crew comfort was eliminated with thoroughness and efficiency which couldn't have been better designed by the enemy. First, all B-25s delivered to this group by the depot had had the hot air heaters removed. This was presumably to decrease the danger of fire in the event of a hit by bullets or flak. Congealed crew members returning from a particularly gelid mission have been heard to wish wistfully that they'd been shot down in flames. Second, B-25s were designed to fly at the warmer medium altitudes and thus the crews did not require electrically heated flying suits, in the opinion of those responsible for equipping said crew members. Leather, sheepskin lined flying clothes were available. In the opinion of most of us, these served admirably to keep the cold inside the garments. This was especially true of the boots. We were surprised, when we removed them after landing, to find that feet weren't solidly enc in ice. Add to this discomfort the Mae West and the flak armor and the result was about a immobile as any combat-ready medieval knight.

One cold winter morning, I walked into Group Operations to give Major Wells, the Assistant Group Operations Officer any assistance he might need. George was responsib for planning and briefing the day's mission, which happened to involve bombing a bridge in Yugoslavia. He also happened to be scheduled to lead the mission. George and Fred Dy the other assistant, led the group in the number of missions and this was the first time the group was to fly over that country. Being first over a new country entitled one to a small degree of bragging rights, so I was mildly shocked when George offered to step aside to l me lead the mission. The surprise was instantly followed by suspicion. A short investigati revealed the motive. The upper air charts showed a free air temperature of -50 degrees. "Nice try, George." About that time Fred came in and went through the identical routine. somewhat resigned Wells was about to accept the inevitable when in walked Lt. Colonel Earl Young. Earl was temporarily assigned to a headquarters with a fighter defense missic that was no longer required, and had wangled an assignment to combat operations. He ha flown some missions with us and, being an experienced officer, was perfectly capable of leading a formation. Three pairs of eyes met. Communication was perfect. I barely had th invitation out of my mouth when he accepted with an alacrity and gratitude which, fleetir made me feel a little guilty.

The poor guy returned about four hours later, and spent about that much time thawing by wrapping himself around the red hot stove in Operations. He didn't seem to even notic the pungent odor of burning clothes or show any inclination to move away from the stov We've been in touch many times for over 42 or so years and he has yet to fail to bring the matter up. Sad to say, it seems that people's values seem to change. He no longer seems grateful for that opportunity to lead the formation.

By then Colonel Chapman had concluded that, regulations or not, the crews - in partic the bombardiers - could not operate at anywhere near peak efficiency under such conditic Somehow he arranged to obtain enough heaters to install in the lead aircraft. As might be expected, bombing accuracy did indeed improve. Equally predictable was the fact that no unexplainable instances of aircraft fires occurred, and, as far as I know, no one in the responsible area of authority ever showed the slightest interest in the subject of unauthori aircraft heaters."

" *And while we're on the subject, let's give the Red Cross a great big hand. We got our first regular installation at Pompeii and it has been a permanent institution every since.*

With bombs away, it's 'So long, flak; hello, donuts and coffee.' And were they life savers after those long, paralyzingly cold rides of last spring! And the generous portions of smiles, conversation and wise cracks served along with the coffee and sinkers, by three dainty bits of American femininity to pep up the morale no end. "

-Major Leonard Kaufmann, Jr.
Commanding Officer of
489th Bombardment Squadron

"Every time Col. Cathcart increased the number of missions and returned Hungry Joe to combat duty, the nightmares stopped and Hungry Joe settled down into a normal state of terror with a smile of relief. Yossarian read Hungry Joe's face like a headline. It was good when Hungry Joe looked bad and terrible when Hungry Joe looked good." (p.54)

"Hungry Joe was a jumpy, emaciated wretch with a fleshless face of dingy skin and bone and twitching veins squirming subcutaneously in the black hollows behind his eyes like severed sections of snake. (He) sprang from spot to spot fanatically with an intricate black camera with which he was always trying to take pictures of naked girls. He could never decide whether to furgle them or photograph them, for he had found it impossible to do both simultaneously. In fact, he was finding it all impossible to do either." (p.52)

HUNGRY JOE

LOOSELY   BASED UPON ➤

Lieutenant Joseph Chrenko
Pilot
340th Bomb Group

HELLER: (Just) Hungry Joe. His real name is Joe Chrenko and he now (1975) is an insurance agent in N.J.

PLAYBOY: Hungry Joe is the one who has screaming nightmares in his tent. Did Chrenko also run around Rome claiming to be a *Life* photographer so he could take pictures of naked girls?

HELLER: Only once.

PLAYBOY: Did he complain about the way you portrayed him in the book?

HELLER: His only complaint is that I didn't use his last name. He feels it would have helped his insurance business.*

t to right: Lt. Clifford (P), Lt. Chrenko (CP), S/Sgt Rackmyer (TG), Lt. Heller (BN), Zaboly (X), S/Sgt Schroeder (RG), and S/Sgt Ryba (G). A 488ᵗʰ crew. June, 1944

(Here we have Hungry Joe and Yossarian on the same mission.)

ght photo: From 488th Bombardment Squadron Yearbook
ft photo: War Diary for the 488th Bomb Squadron, June 1944

Heller's 7th mission. To Cecina, Italy.

xcerpt from a *Playboy* Magazine interview of Joseph Heller, June 1975

"All right, I'll dance with you," she said, before Yossarian could even speak.
"But I won't let you sleep with me."
"Who asked you?" Yossarian asked her.
"You don't want to sleep with me?" she exclaimed with surprise.
"I don't want to dance with you." (p.152)

Luciana

"Luciana fled mirthfully along the sidewalk in her high white wedgies..."
(p.161)

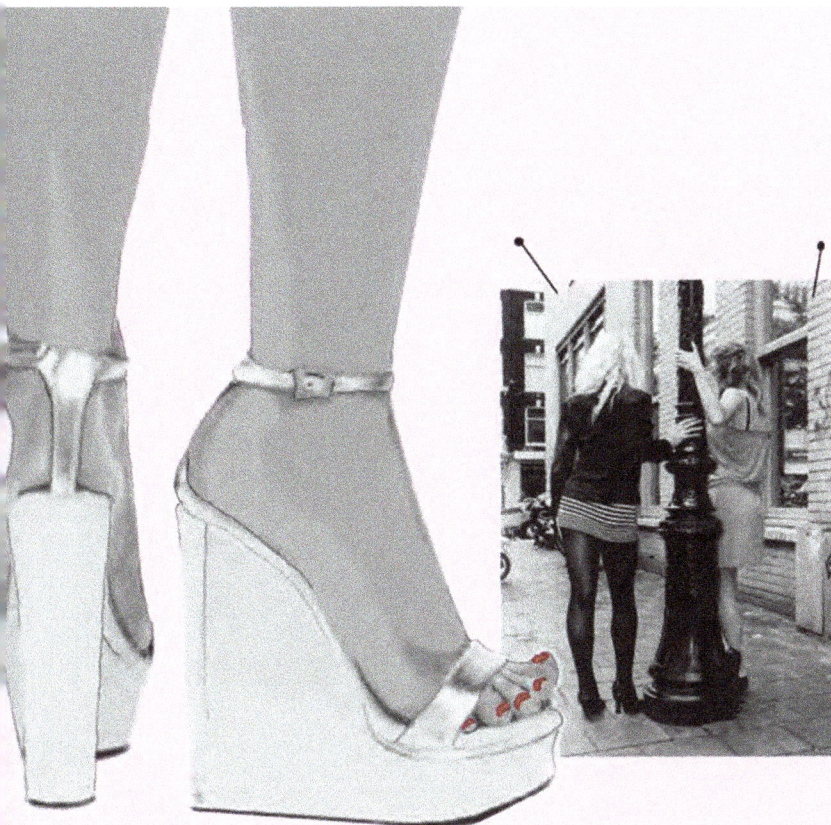

The True Story

- Joseph Heller

"Luciana was Yossarian's vision of
a perfect relationship. That's why
he saw her only once and perhaps
that's why I saw her only once. If
he examined perfection too closely,
imperfections would show up."

"... and the men scrambled out (of the crippled and sinking airplane) as speedily as they could in their flaccid orange Mae West life jackets that failed to inflate and dangled limp and useless around their necks and waists. The life jackets failed to inflate because Milo had removed the twin carbon-dioxide cylinders from the inflating chambers to make the strawberry and crushed-pinapple ice-cream sodas he served in the officer's mess hall and had replaced them with mimeographed notes that read: 'What's good for M&M Enterprises is good for the country.' (p.301)

"But I make a profit of three and a quarter cents an egg by selling them for four and a quarter cents an egg to the people in Malta I buy them from for seven cents an egg." (p.226)

7¢

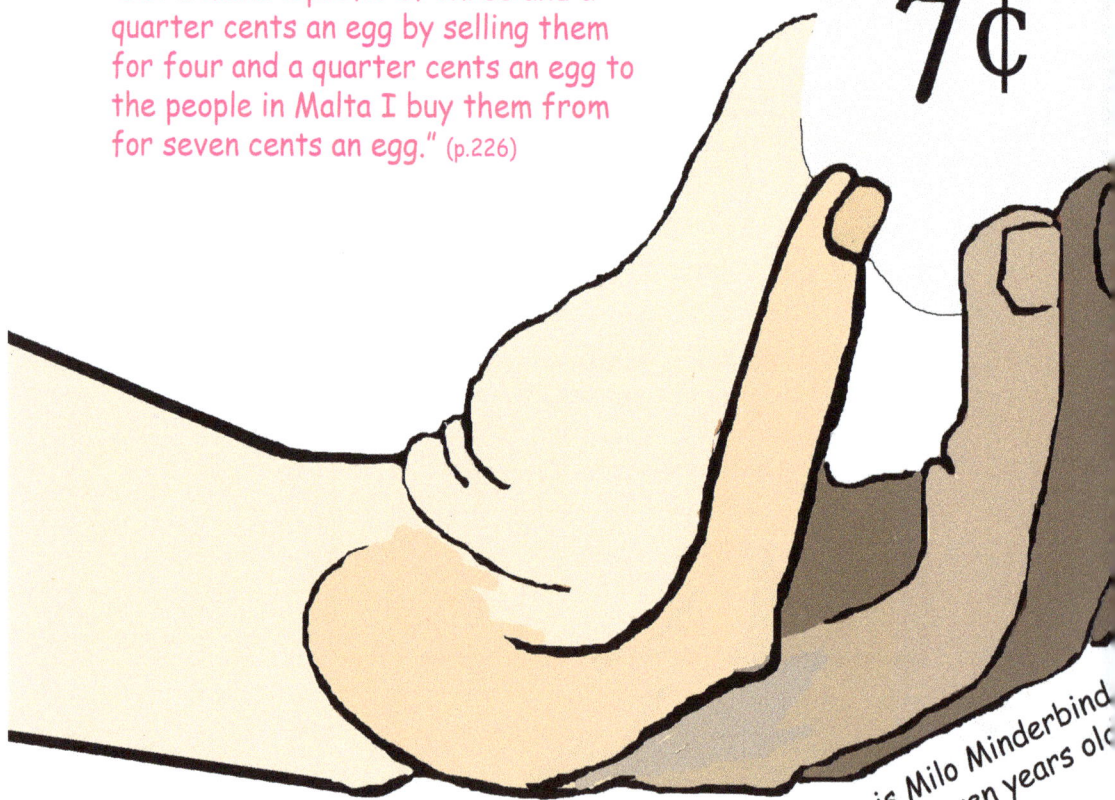

"My name is Milo Minderbind, I am twenty-seven years old

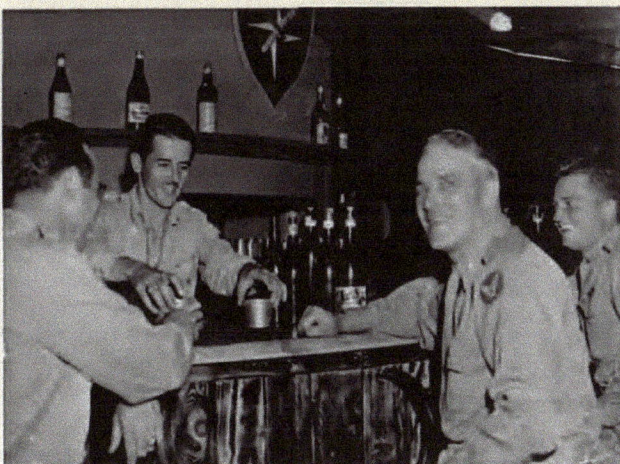

ajor Benjamin Kanowsky
Pilot and Mess Officer
340th Bomb Group

Decorations
*"Couple of Distinguished Flying
Crosses and several Air Medals"*

L to R: Bill Chapman (Cathcart), Ben Kanowsky (Minderbinder),
Bob Knapp (Dreedle)

The True Story
- Ben Kanowski

"That Mae West story in the book, for example. It was true. We used our shell cartridges to store the beer. We improvised a lot. Col. Chapman was invited to a top secret Pentagon briefing a little while ago. All the top brass were there. Admirals, Generals, Defense Dept. officials. They all went down to the sub-basements. The colonel said he never saw so many guards before. At the meeting they showed them some new rockets and missiles. It was top secret. Chapman looks at one of those things, and then stands up to address the whole group. He's the lowest ranking officer there, remember.

Pointing to the rocket, he says, 'One of my men made that rocket in 1944.' When the lights came on everyone was gone. 'That was another reason I was a colonel so damned long.'

Anyway, I built that rocket. When I made it, it had tinsel, chaff, tin foil in it. It was to be placed under our bomber wings and dropped so that all that metal would mess up the enemy's radar. Chapman wants the sketches and drawings I have so he can prove it to the Pentagon.

When I was testing the rocket, I rigged it up on a wire from my tent to a tree a distance away. On the first test trial, that thing flew out and almost knocked Major Major's head off. They all said I tried to kill him. Shit, I was inside the tent and didn't even see him"

Catch-22 Characters Reunite in Albuquerque. Milo Minderbinder/Ben Kanowsky

"CHIEF - could barely read or write and had been assigned to Captain Black as assistant intelligence officer. " (p.43)

"HAVERMEYER - a lead bombardier who never missed. " (p.29)

"He was a glowering, vengeful, disillusioned Indian who hated foreigners with names like Cathcart, Korn, Black and Havermeyer and wished they'd all go back to where their lousy ancestors had come from."

" ... held ma men rig six plane steady and as sit ducks w he follo the bo all the down thro the plexig nose with c interest gave the Gern gunners belo the time t neede

CHIEF WHITE HALFOAT / HAVERMEYER

LOOSELY BASED UPON ➤

Wells, Myers, and Helfertch.
Wells and Helfertch each flew
their 100th mission together.

Major Vincent Myers
Group Bombing Officer
Bombardier, 340th Bomb Group

The True Story
- George Wells

"I have known a few exceptional bombardiers but none were in the same league with Vincent Myers, better known to us as "Chief" Myers. There was a bombardier who seemed to have no equal. With the conversion to the Norden bombsight early in 1944, we increasingly took on the more difficult 'pinpoint' targets rather than the easier to hit 'area' targets. It was during this transition Chief's real capabilities surfaced. He became the leader of the lead bombardiers. On one series of thirteen missions together, he, on each one of them, got direct hits on the most difficult of the pin-point targets - the twenty foot wide rail ridge from over two miles away. He was just amazing in his ablity to hit a target even when the defenses were intense. He was fearless without equal, this does not imply he was recklessly brave - he just had more calm fighting ability than the rest of us.

He was a very human person who got fun out of life. Once things were a little duller than usual on a stand-down day due to weather and Chief talked Fred Dyer and Cal Moody and me into having a flare gun battle with the British Army Liaison Officer and his assistants. Have you ever seen a flare hit a tent? To say the least - we were lucky we didn't start a fire we couldn't control or hurt someone. It was also fortunate Chapman was gone for the day.

On another occasion, after having arranged for the British Liaison Officer to go on his first combat mission a 'milk run', Chief and I took down the Britisher's tent, while he was on the mission and marked his gear for 'Next of Kin'. Thank heavens he returned safely. We did help him put his things back together again."

On occasion Heller filtered through a crewmates personality and make-up and creatively developed two distinct factitious characters. This happened in the case of Vincent Myers. This bombardier split, atom-like, into two of <u>Catch-22</u>'s most recognizable characters - the over-the-edge, loose cannon, Chief White Half Oat and ice-ace Havermyer.

The True Story
- George Wells

Chief White Halfoat was out to revenge himself upon the white man. He could barely read or write and had been assigned to Captain Black as Assistant Intelligence Officer.

> Handsome, swarthy, Indian from Oklahoma with a heavy, hard-boned face and tousled black hair. A half-blooded Creek from Enid. (p.43)

"Again, *this is* Chief Myers. He was every one of those words except he was a half-blooded Comanche from Apache, Oklahoma.

Chief did have an incident with Gen. Knapp's MOE (a lieutenant) at a rest hotel on the northeast coast of Corsica on a Saturday night. Chief's notoriety as a Golden Gloves champ was well known. Sometimes a man likes to show his masculinity by talking tough to such a person as Chief. There was a small band playing music on this Saturday night in late 1944 at the hotel. The aide pushed Chief too far and Chief hit him, picked him up and tossed him into the musician's drum. This may be the connection to Halfoat, on many occasions, hitting Gen. Dreedle's son-in-law, Col. Moodus (Counterpart General Knapp's son, Robert Knapp, Jr., also was Gen. Knapp's aide.) Again, Heller created a name, Col. Moodus, most likely drawing upon one close to that of a 340th officer, Cal Moody, the group navigation officer.

There is no doubt Havermyer is Capt. Vincent Myers. *Catch-22's* description of him: Lead bombardier. Best one in the group. Flew straight and level over the target. Never missed. Used candy to bait mice and then shot them. But Heller didn't make the character an Indian. Heller used Chief Myer's Indian background in the character Chief White Halfoat. Chief and I flew a lot of missions together as lead pilot and lead bombardier."

"Now when a beautiful perfect echelon, or even the
whole damn formation, comes over low enough
to blow the sugar off the Red Cross donuts,
we know without listening to the radio that
they're saying,
'Mission completed, bridge finito.'

-Major Leonard Kaufmann, Jr.
Commanding Officer of
489th Bombardment Squadron

"You mean I can't
shoot anyone I
want to?"

"I guess you think you're pretty goddamn smart, don't you?"
General Dreedle lashed out at Colonel Moodus suddenly.
Colonel Moodus turned crimson again. "No, Dad, it isn't ..." (p.312)

COLONEL MOODUS

LOOSELY   BASED UPON ➤

Bob Knapp, Jr. and father. 1931. Airplane - 0A4

Father and son in combat together.
Corsica, 1944

2nd Lt. Robert D. Knapp, Jr.
Aide to B/G Robert D. Knapp
321st Bomb Group

The True Story
about
Robert D. Knapp, Jr.

pl. Robert D. Knapp, Jr., 22, an aerial gunner in a B-25 Group, is the son of the
ing's commanding general, Brig. Gen. Robert D. Knapp. In response to a reporter's
estion, he replied that the tangible rewards for such rank in the family are small.
bout once a week the old man has me up to eat with him. He has pretty good chow.
e doesn't like candy and he gives me his rations; I suppose he would give them to
e if I were only a sergeant

Shortly after he arrived at his unit, his father accompanied him out to this new
oup."I want to see how you report to an officer." Cpl. Knapp replied, "I imagine
ve reported to more officers than you have." Whereupon he went through the Army
tual in a very GI manner, pleasing his father a great deal. He saluted every officer
the area.

He had just begun to unpack when a gunner with about 59 missions came in to
s tent. "Listen, new boy, you are not in the States. This is your war. They are
ooting at you. This outfit doesn't hold contests for saluting. We salute when we
ve to and let it go at that." Knapp did not say anything.*

* *Men of the 57th*, March 1982.
With a 12th AAF B-25 Wing 1944

ater Bob, Jr., as a new 2nd Lieutenant, became his father's Aide.)

Henry Fonda

"Major Major Major Major had had a difficult time from the start ..."

"Major Major had been born too late and too mediocre. Some men are born mediocre, some men achieve mediocrity and some men have mediocrity thrust upon them. With Major Major it had been all three ..". p.82

"It was still more frustrating to try to appeal directly to Major Major, the long and bony squadron commander, who looked a little bit like Henry Fonda in distress and went jumping out the window of his office each time Yossarian bullied his way past Sergeant Towser to speak to him about it (the dead man in his tent.)" p.22

MAJOR MAJOR MAJOR MAJOR

LOOSELY   BASED UPON ➡

Capt. Randall C. Cassada (back to camera)
Catch-22's Major Major Major Major

L to R-
Gen. Robert D. Knapp
Catch-22's Gen. Dreedle

Lt. Gen. John K. Cannon

Col. Willis F. Chapman
Catch-22's Col. Cathcart

Maj. Joseph W. Ruebel
Catch-22's Maj. Danby

Captain Randall C. Cassada
Squadron Commander
340th Bomb Group

The True Story
- George L. Wells

He is Randall C. Cassada and he was a captain and squadron commander when I arrived in the squadron. Heller says, "Major Major Major Major (the squadron commander) was long and bony - looked a bit like Henry Fonda in distress," (p.22) a description that fits Cassada. And like Major Major Major Major he also played basketball once in a while and was withdrawn by nature. Didn't talk much. In the Group there was also another officer named Major Major, upon whose name Heller drew.

The squadron was, in my opinion, the best in the Group. Most of the people who ended up in key jobs came from the 488th. That movement of key people alone justified the fact that the 488th was the top squadron in the top Group in the top Wing in World War ll. So Cassada must have been doing something right.

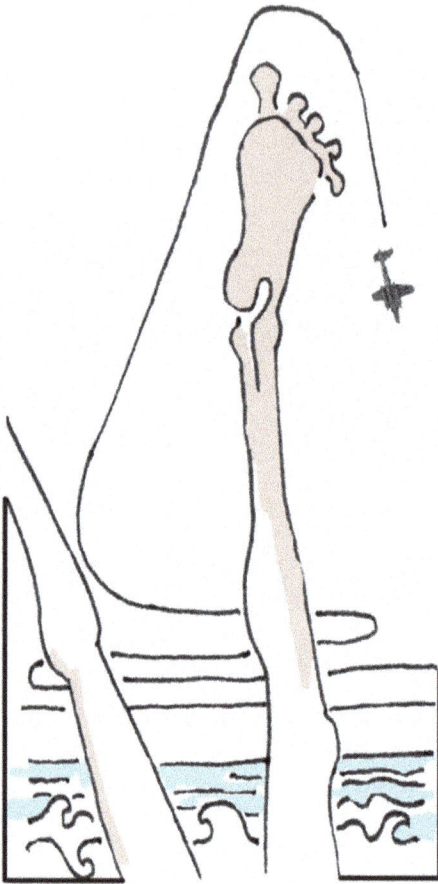

"...with a great growling, clatterin
roar over the bobbing raft on whic
blond pale Kid Sampson, his naked
sides scrawny even from so far aw
leaped clownishly up to touch it at
the exact moment some arbitrary
gust of wind, or minor miscalculati
of McWatt's senses dropped the
speeding plane down just low enoug
for a propeller to slice him half aw

"...and then there were just
Kid Sampson's two pale, skinny
legs, still joined by strings
somehow at the bloody truncated
hips... before they toppled
over backward into the water..." (p.

"...a great, choking moan tore
from Yossarian's throat as
McWatt turned again, dipped
his wings once in salute ...
and flew into a mountain." (p.333)

KID SAMPSON

L O O S E L Y   BASED UPON ➡

1st Lt. William Y. Simpson
Pilot
340th Bomb Group

George Wells & Bill Simpson

"This is the picture of one of our B-25's that was [cu]t by another of our B-25's over the target. Note [th]at the tail gunner's position was cut away with [th]e tail gunner in it." - George

The True Story
- George L. Wells

"Heller isn't writing about a fictitious event. There can be no doubt he is writing about the 340th. Events were there. Things were there and people, changed a bit, were there. In that raid he used the squadron mid air collision over the target at La Spezia and combined it with the bombing of the disabled Italian cruiser which took place Sept. 23, 1944. The mid air collision took place on Jan. 21, 1945.

Heller may have drawn on Simpson, who piloted the plane that cut the tail section off of the other plane, and who was also very young, blond, and scrawny, and made him pay for slicing the tail gunner out of the plane by having him (as Kid Sampson) crash into the mountain."

DOUGLAS ORR

LOOSELY BASED UPON →

Douglas Orr
Bombardier/Navigator
321st Bomb Group

"Can't you just picture him?" he (Chaplain) exclaimed with amazement. "Can't you just picture him in that yellow raft, paddling through
the Straits of Gibraltar at night with that tiny little blue oar."
"With that fishing line trailing out behind him, eating raw codfish
all the way to Sweden, and serving himself tea every afternoon." (p.439)

n early 1943 the men of the 321st had three enemies, Jerry ME 109's, Ack Ack, and the
literranean itself. Returning from a mission in April over the sea, a 447th ship passed over
nvoy and was jumped by a flock of Jerry fighter planes. It didn't take long for the B-25 to
estroyed.

°ilot Albert Duke of Baltimore, Md. held the crippled B-25 steady long enough to crash
on the sea. The plane sank 30 seconds after hitting the water. Bombardier-Navigator
glas Orr of Lakewood, Ohio and copilot James Ackley of Pittsburgh, Pa. had just enough
to toss out a rubber life raft, grab their jungle kits and climb into the raft with the rest of
crew.

After twelve hours of paddling towards Tunisia, guided by Orr, at midnight the raft reached
behind enemy lines. The shipwrecked fliers, still led by Doug Orr, holed up in a tall brush
eep warm and traveled towards US lines. Once they encountered a herd of goats. They
ed some of them and slept with them to keep warm.

After four days of walking, during which they wore out their shoes on the rough terrain, the
made contact with an advance British patrol. Got some shoes and some food. The British
amazed that the men had traveled all that distance without encountering any Nazi patrols.

the four days in
my territory, the
food consumed
some emergency
olate rations, a
stolen eggs, goat's
and a hunk of
d purchased from
uple of Arabs*.

rs *and Stripes*
rinted in
brances, compiled
Lynch of the
mb Wing.)

Orr's raft and blue oar

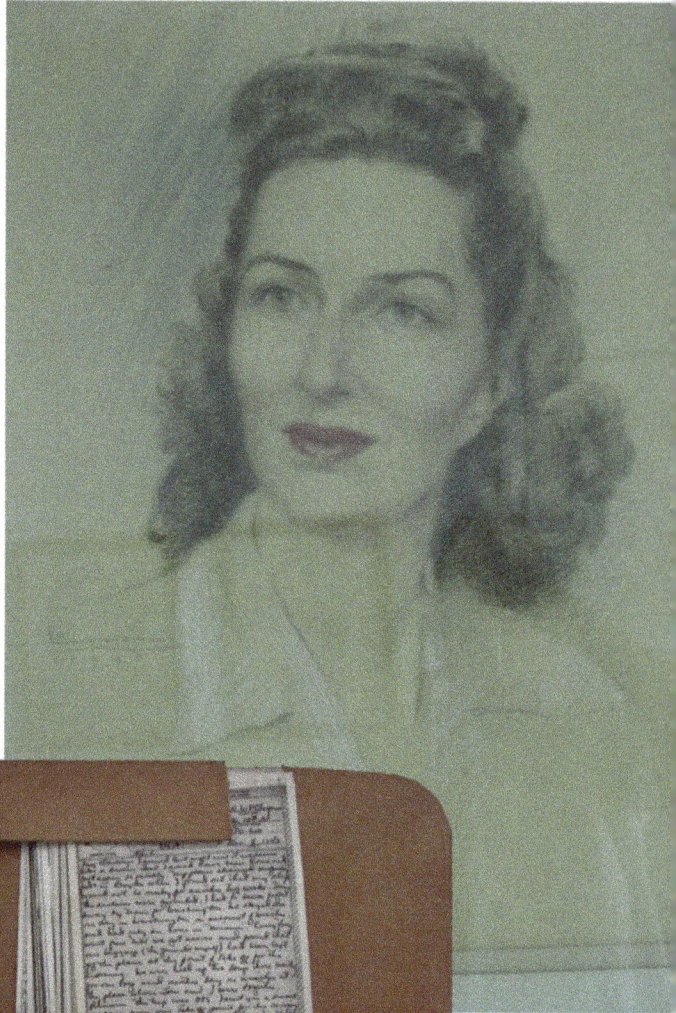

VICTORY MAIL, or V-Mail, was the result of a process of microfilming letters during WWII onto special sheets to reduce the mail bulk taking up valuable cargo space needed for war materials.

Procedure:
1. Put letters on film which saved 98% of cargo weight and space.
2. Send film overseas.
3. Soldiers/sailors enlarge and print letters on long rolls of paper.
4. Each print of a letter, 41/2x51/2 inches, is cut apart by hand, folded and placed in an envelope and delivered.
5. Military personal were issued 2 sheets of V-Mail stationary a day.

"All the officer patients in the ward were forced to censor letters written by all the enlisted-men patients ...

Yossarian was disappointed to learn that the lives of enlisted men were only slightly more interesting then the lives of the officers." *(p.8)*

1944-1945

Examples of V-Mails sent by Bill Chapman to his wife, Charlotte, who preserved them in this small, soft leather folio.

(See p. 123)

CATCH-22'S WHO'S WHO

JOHN YOSSARIAN

He was the best man in the group at evasive action, but had no idea why.

About Joseph Heller:

Lt. Joseph Heller - "A nearby tent . . . was the tent of a friend, Francis Yohannon, and it was from him that I, nine years later, derived the unconventional name for the heretical Yossarian. The rest of Yossarian is the incarnation of a wish."[1]

"At a reunion years later," recounts George Wells, "Francis Yohannon told us that Heller had used his name for the character Yossarian. Heller used the first part of his name and combined the last part with the fact that he was, by birth, Syrian."[2] (*Catch-22*'s Yossarian was Assyrian.)

Captain John Yossarian sprang from the mind of Joseph Heller. He took with him Heller's thoughts, his mannerisms, his creativity and unconventionality and, perhaps, even his appearance. They shared a devil-may-care attitude, a logical outlook, a quirky nature, and an irreverent, quick humor. Indeed, Joseph Heller birthed Yossarian.

In an article that she published for The New York Times Book Review, writer Barbara Gelb once described her friend, Joseph Heller. "For twenty years now, I have managed to overlook his frequent sulkiness, his gluttonous table manners and his tendency to growl 'No' before he even knows what the question is. I have stayed on good terms with him largely because I relish his aberrant sense of humor and his skewed way of looking at life—an outlook he insists has changed little since he wrote *Catch-22*."

GENERAL DREEDLE
Wing commander with nubile nurse and captive son-in-law.

About General Robert D. Knapp:

Brigadier General Robert D. Knapp was, indeed, older than his men. He was born in 1897; his pilot's license #185 was signed, years later, by Orville Wright who, with his brother Wilber, lived three houses away from young Knapp for a few months while working on their fledging airplane. He served in World War I, he flew in the border war against Pancho Villa, he barnstormed across the country trying to get Americans interested in aviation, he flew the first airmail route from Montgomery, Alabama and New Orleans, Louisiana. These events occurred before most of the men he commanded in WWII had even been born.

Knapp himself was a man who was born to fly. There was nothing he would rather do. As well, this was a man who was born to lead. Yes, he had spent much of his life doing his job well. What greater confirmation than to be so highly respected by the men under his command, "his boys."

Knapp had a son, Robert D. Knapp, Jr. who also entered WWII. When Knapp, Jr. became a second lieutenant he also became his father's aide de camp. Patterning on Lt. Knapp, Jr., Heller loosed his intense creativity in developing Colonel Moodus, changing him into a son-in-law in *Catch-22* who General Dreedle delighted in keeping miserable. Here is where Heller and Knapp parted company. General Knapp and his son stayed on the serious course of dealing with this war. General Knapp had become a bit of a legend. His life was a history book for facts, a Jack London novel for adventure, and a dash of Will Rogers for easy humor.

COLONEL CATHCART

Bomb Group commander who calculated day and night in service of himself.

Colonel Cathcart was a very large, pouting, broad-shouldered man with close-cropped curly dark hair that was graying at the tips and an ornate cigarette holder that he purchased the day before he arrived in Pianosa to take command of his group. He displayed the cigarette holder grandly on every occasion and had learned to manipulate it adroitly. Unwittingly, he had discovered deep within himself a fertile aptitude for smoking with a cigarette holder. As far as he could tell, his was the only cigarette holder in the whole Mediterranean theater of operations, and the thought was both flattering and disquieting. *(p. 185)*

About Colonel Willis F. Chapman:

Here's what George had to say about Colonel Cathcart and Colonel Chapman. "Heller made many traits, appearance and personality comments about Cathcart, the character he drew from Colonel Chapman—some are very true and some are very wrong. But the one that fits him the most is the one about Col. Chapman when he walks up to the front of the room on his first meeting with the combat crews and told us we were now going to "fly straight and level" for some minutes from the Initial Point to the target, and in his hand was this long Zeus cigarette holder. I didn't smoke so I wasn't pleased with this new colonel. Well, that colonel went on to become a great leader and made the 340th Group outstanding. It became a leader in hitting pinpoint targets. His several-minute bomb run, straight and level, gave the bombardiers (even average ones) time to determine wind conditions, find the target, set the sight, and hit the target. Some of Heller's words are right on (the good); others are completely wrong. Colonel Chapman was a top tactician with a very innovative mind. He took us from being an average group to one that could fly great formations with a large number of planes, one that could get the combat job done without problems, one with ground crews as well as aircrews doing the job successfully without complaint. The people understood the need and did all the jobs gladly, with competence, thanks to Colonel Chapman. I really don't see many incidents apply to Colonel Chapman in Heller's book, but the descriptions fit so correctly on a number of things like size, cigarette holder, and a number of the positive adjectives."[3]

One morning at Group breakfast, immediately after the new commander's arrival, the 340th HQ staff all popped out Colonel Chapman style cigarette holders that they had flown in from Rome. Deputy Group CO Lt. Col. "Mac" Bailey and voluptuous Red Cross gal Dita Davis instigated the joke

Chapman

Cartoon of Chapman with cigarette holder
drawn by Captain Jack C. Mair, pilot, 489th.

MAJOR ___ DE COVERLEY

Everyone was afraid of him and no one knew why.

Each time the fall of a city like Naples, Rome or Florence seemed imminent, Major __de Coverley would pack his musette bag, commandeer an airplane and a pilot, and have himself flown away, accomplishing all this without uttering a word, by the sheer force of his solemn, domineering visage and the peremptory gestures of his wrinkled finger. A day or two after the city fell, he would be back with leases on two large and luxurious apartments there, one for the officers and one for the enlisted men, both already staffed with competent, jolly cooks and maids. A few days after that, newspapers would appear throughout the world with photographs of the first American soldiers bludgeoning their way into the shattered city through rubble and smoke. Inevitably, Major __de Coverley was among them, seated straight as a ramrod in a jeep he had obtained from somewhere, glancing neither right nor left as the artillery fire burst about his invincible head and lithe young infantrymen with carbines went loping up along the sidewalks in the shelter of burning buildings or fell dead in doorways. He seemed eternally indestructible as he sat there surrounded by danger, his features molded firmly into that same fierce, regal, just and forbidding countenance which was recognized and revered by every man in the squadron. *(pp. 130-131)*

About Major Charles Jerre Cover:

About the actual event, Heller relates: "On June 4, 1944, the first American soldiers entered Rome. And no more than half a step behind them, I think, must have come our own squadron's resourceful executive officer, for we received both important news flashes simultaneously: the Allies had taken Rome, and our squadron had leased two large apartments there, one with five rooms for the officers and one with about fifteen rooms for the enlisted men. Both were staffed with maids, and the enlisted men, who brought their food rations with them, had women to cook their meals."[4]

George gives us his view on "de Coverley." "Now here is a good example of copying name and actual actions by a real person in the book. The 488th squadron executive officer was Major Charles J. Cover. You can't get much closer in a name and position and rank and still call it fictitious. Cover was medium height, not tall. His face could be called craggy as in *Catch-22*. Cover was a fine older gentleman to us people in the squadron. (Cover was born in 1896 and would have been all of 48 years old.) Very well liked. He had dignity but was not fierce. He seemed old to us, was splendid, slightly stern face and maybe a slightly patriarchal face. I recall that he did play a lot of horseshoes. He did arrange for barbers who were from Italy. He had arranged for an apartment in Italy to be used by the enlisted men on short rest trips. He was well liked by members of the squadron."[5]

CAPTAIN PILTCHARD AND CAPTAIN WREN

Joint squadron operations officers who flew endlessly and fearlessly.

Captain Piltchard and Captain Wren, the inoffensive joint squadron operations officers, were both mild, soft-spoken men of less than middle height who enjoyed flying combat missions and begged nothing more of life and Colonel Cathcart than the opportunity to continue flying them. They had flown hundreds of combat missions and wanted to fly hundreds more. Nothing so wonderful as war had ever happened to them before, and they were afraid it never would happen again. (p. 144)

About Capt. Fred W. Dyer and Capt. George L. Wells:

About this reference, George says: "Dyer and I were both captains until late 1944. Dyer was Piltchard and I was Wren. Less than middle height (Dyer was 5'9" and I was 5'7") we both did like to fly a lot and we became better by flying more. Dyer and I would go up and fly in two planes on each other's wing. Many times on down days or when we weren't on a mission, we would fly in heavy weather conditions in formation- practice single engine in formation. We flew the planes in practice under various types of conditions to fully understand its performance as well as its limitations. We really set the tone for combat in the 340th. What we both did was to (1) eliminate the basis for bellyaching by the combat crews due to any concern over the B-25's performance even when heavily damaged or due to fear for oneself while in combat and (2) set an example by flying additional missions without having to do it—and never picking a mission because it was considered a milk run; in fact, we did the opposite – the more challenging [the mission] the more we competed with each other."[6]

Maj. Fred Dyer (9Z), who was a stand-by pilot on the Hornet for Doolittle's Tokyo raid, Lt. Col. Lou Keller (6P) and Maj. George Wells (8R) after practicing single-engine maneuvers in the B-25.

"Fred, Lou and I flew this flight to show the combat crews that the new H model B-25s could fly and land on a single engine, which was of major importance. Note we each have a prop at standstill." - George Wells

CHAPLAIN A. T. TAPPMAN

Everyone was always very friendly toward him, and no one was ever very nice; everyone spoke to him, and no one ever said anything.

It was love at first sight. The first time Yossarian saw the chaplain he fell in love with him. *(p. 7)*

About Chaplain James H. Cooper:

George: Of course Chaplain Tappman is a comical joining of title with a name. The real chaplain was James Cooper. There is a tie in Heller's use of the name Tappman for a Group level officer to the name Chapman also at Group level,"[7] says George Wells. Fictional Chaplain Tappman and counterpart Chaplain Cooper both lived alone in a spacious tent in the woods a goodly distance from the squadron encampments and Group Headquarters.

THE SOLDIER IN WHITE

No sound at all came from the soldier in white all the time he was there.

The people got sicker and sicker the deeper [they] moved into combat, until finally in the hospital that last time there had been the soldier in white who could not have been any sicker without being dead, and he soon was. (p. 165)

GENERAL ROBERT KNAPP AND HIS 'SOLDIERS IN WHITE'

A bombardier in the 489th, Hal Lynch, witnessed an event so stirring that it imbedded itself deeply within his memory and remained with him for the remainder of his long life. In 2002 he transferred this experience to paper as he penned the following.

" During World War II my B-25 medium bomber group was stationed on the beautiful Island of Corsica, a French possession in the Tyrrhenian Sea. Our dirt runway bordered the sea. On a clear day the island of Elba to the east was clearly visible. In October of 1944 I witnessed a most remarkable scene from the center of that dirt runway. It happened to be a special occasion, the recognition of our group for its year of service in combat in the European Theater of Operations. During this year many men in the group had been killed or were interred in Italy or had gone down at sea.

Thus a memorial service was planned. A temporary platform, facing the sea, had been constructed on the runway. Dignitaries from the 12th Air Force and from the United States Congress had been invited to be a part of the tribute. Our 57th Bomb Wing Commander, Brigadier General Robert D. Knapp, had been asked to serve as the main speaker.

On that memorable day in 1944 a 12th Air Force general opened the program with a short speech, given while facing the dignitaries, with his back to the Tyrrhenian Sea.

And then General Knapp was introduced.

He walked to the microphone, glanced at the dignitaries, and then turned his back to them and faced the sea. For the next twenty minutes the general spoke to his men, the men who had been lost in Italy or in the sea. He apologized to them for his part in sending them into combat – for not being a better leader, a closer friend. And then in a throbbing voice, he told the men he would never forget them and he would never allow our nation to forget them – that he would try to repay them for their great sacrifice by living beyond himself in the years ahead.

The general never did look back at the dignitaries on the platform. In fact, he had probably forgotten they were there. This was a private moment. He was alone with his men. When he completed his tribute he walked away from the platform, heading in the direction of his favorite B-25 parked at the end of the runway.

I realized that I had been part of an unforgettable experience. I had always accepted the fact that General Knapp was a great leader, but on that beautiful day on Corsica I realized that he was more than a great leader. This outstanding man was also an inspiration to all who knew him."

Men of the 57th Bomb Wing, Fall, 2002, Vol. XXXV No. 3 p

LUCIANA

She was a tall, earthy, exuberant girl with long hair and a pretty face, a buxom, delightful, flirtatious girl. *(p.152)*

Joseph Heller says:

"Luciana is just what happened to me in Rome. Luciana was Yossarian's vision of a perfect relationship. That's why he saw her only once, and perhaps that's why I saw her only once. If he examined perfection too closely, imperfections would show up."[14]

DOC DANEEKA

Doc Daneeka had been drafted and shipped to Pianosa as a flight surgeon even though he was terrified of flying. He felt imprisoned in an airplane. In an airplane there was absolutely no place in the world to go except to another part of the airplane . . . (p. 41)

"You're dead, sir."

"It's true, sir," said one of the enlisted men. "The records show that you went up in McWatt's plane to collect some flight time. You didn't come down in a parachute, so you must have been killed in the crash."

"Gee, I guess he really is dead," grieved one of his enlisted men in a low, respectful voice (still standing next to Doc). "I'm going to miss him. He was a pretty wonderful guy, wasn't he?"

"Yeah, he sure was," mourned the other. "But I'm glad the little fuck is gone. I was getting sick and tired of taking his blood pressure all the time." (p. 335)

The military doctors drew flight pay by logging in flight time, often in the night hours. As the war progressed the rules were bent a bit with the realization that this was too dangerous to expose doctors to when their skills were so needed on the ground. So their names were sometimes logged in on a flight that their bodies did not accompany. Joseph Heller delightedly embellished on that fact with this episode of Doc supposedly having been killed in McWatt's plane.

About Captain Benjamin J. Marino, M.D.:

George: Our flight surgeon was Captain Benjamin W. Marino. Heller's description of him is close. He did seem sad and was a recluse. We used to say when you went to see him all you got was an aspirin. Most people can adapt to the situations they find themselves in during their life, but Dr. Marino couldn't find much in his daily life, during the war, to please him[8]

"He was compassionate and caring. The men laughingly swore that Marino delivered more Italian babies than the number of soldiers he cared for."[9]

MAJOR DANBY

"It must be nice to live like a vegetable, he conceded wistfully."
"A cucumber or a carrot."

. . . the gentle, moral, middle-aged idealist.

. . . "I'm a university professor with a highly developed sense of right and wrong, and I wouldn't try to deceive you. I wouldn't lie to anyone."

"What would you do if one of the men in this group asked you about this conversation?" [Yossarian asked.]

"I'd lie to him." *(p. 434)*

About Major Joseph W. Ruebel:

George: That's Joseph W. Ruebel, our Group Operations Officer. Chief, Dyer and I worked for him; he was our immediate boss. Joe Ruebel was a great boss with a great personality. Everyone liked him. He was thorough at his job and was a gentle and modest man. He did let us three influence combat tactics decisions because all three of us had so much more combat experience then he did. Chief and I tented with Joe when we moved up to Group Headquarters. [11]

HUNGRY JOE

He crumbled promptly into ruin every time he finished another tour of duty... he began screaming in his sleep. (p.53)

. . . Every time Colonel Cathcart increased the number of missions and returned Hungry Joe to combat duty, the nightmares stopped and Hungry Joe settled down into a normal state of terror with a smile of relief. *(p.54)*

About Joseph Chrenko, pilot:

Heller: "In Yohannon's tent also lived Joe Chrenko, a pilot I was especially friendly with who later, in several skimpy ways, served as the basis for the character Hungry Joe. In that tent with them was a pet dog Yohannon had purchased in Rome. In my novel I turned the dog into a cat to protect its identity."[12]

Heller: [Just] Hungry Joe. His real name is Joe Chrenko and he now [1975] is an insurance agent in New Jersey.

Playboy: Hungry Joe is the one who has screaming nightmares in his tent. Did Chrenko also run around Rome claiming to be a Life photographer so he could take pictures of naked girls?

Heller: Only once.

Playboy: Did he complain about the way you portrayed him in the book?

Heller: His only complaint is that I didn't use his last name. He feels it would have helped his insurance business.[13]

MILO MINDERBINDER

Milo had been caught red-handed in the act of plundering his countrymen, and, as a result, his stock had never been higher.

"What's your name, son?" asked Major ___ de Coverley.
"My name is Milo Minderbinder, sir. I am twenty-seven years old."
"You're a good mess officer, Milo."
"I'm not the mess officer, sir."
"You're a good mess officer, Milo."
"Thank you, sir, I'll do everything in my power to be a good mess officer." *(p.134)*

About Benjamin Kanowsky, Pilot & Mess Officer:

George: Ben Kanowsky was the Mess Officer whom the 340th members considered the source of Heller's fictitious Milo. His function was to try to obtain, by trade or purchase, food (greens, eggs, etc.) for our enlisted and officers' mess tents. Heller took gross liberties with this character. Ben also had an inventive mind and did work on some ideas for the group. One of these ideas involved firing chaff ahead of the element to increase defense against ack-ack (flak), anti-aircraft artillery.[15]

Joe Heller, however, noted, "Here we may be a little close even for my comfort, both because of a slight similarity in names and because of the activities and opportunities common to all mess officers. The name of my mess officer was, believe it or not, Mauno Lindholm. Like other mess officers on Corsica, he had a certain amount of money at his disposal to purchase fresh provisions from local sources. He flew to other places regularly for fresh eggs, meat, and vegetables. Whether he used his position and resources to make money in the black market is his own secret; there was no reason to believe he did not, and much reason to believe he did. Like Milo, Mauno, if memory serves me correctly, had a mustache. From here on, all similarities end. Milo's moral rationalizations are all fictional, and everything that happens to him are but extensions of the possible into the fantastic.

HAVERMEYER / CHIEF WHITE HALFOAT

He had grown very proficient at shooting field mice at night with the gun he had stolen
from the dead man in Yossarian's tent.
His bait was a bar of candy ... (p.30)

Havermeyer was the best damned bombardier they had, but he flew straight and level all the way from the IP to the target, and even far beyond the target until he saw the falling bombs strike ground and explode in a darting spurt of abrupt orange that flashed beneath the swirling pall of smoke and pulverized debris geysering up wildly in huge, rolling waves of gray and black. *(p. 29)*

About Major Vincent Myers:

George: "There is no doubt Havermeyer is Capt. Vincent Myers ("Chief" Myers). *Catch-22*'s description of him: Lead bombardier. Best one in the group. Flew straight and level over the target, never missed, used candy to bait mice and then shot them. But Heller didn't make the character an Indian. Heller used Chief Myers' Indian background in the character Chief White Halfoat. Chief and I flew a lot of missions together as lead pilot and lead bombardier. Chief and I were the best of friends and always flew together whenever we could. We tented together for over a year in combat so I knew him as well as or better than anyone else in the group. If there was a tough target or one hard to find we, as a team, flew a large number of them. Chief was a gentleman and a great and loyal friend, father, and husband during and after the war. Chief was the best bombardier in the Group by far, and the bravest bombardier I ever knew. In fact, in my opinion, I think he was one of, if not the, best in WWII. When the Norden bombsight (excellent for pinpoint targets such as bridges) replaced the Mark IX bombsight (good for area targets but not pinpoint), orders were given for the 5-minute bomb run. Straight and level with no evasive action. No one ever took Chief to task, as mentioned in *Catch-22* for flying straight and level over the target, but he would want us to hold it a few seconds longer after Bombs Away before the break away to give our vertical cameras a better chance of getting a picture of the target area for results.

Chief did shoot at mice a couple of times while in the 488th to try to keep them away from his candy ration, and he was a good shot."[16]

Chief White Halfoat was out to revenge himself upon the white man. A handsome, swarthy, Indian from Oklahoma with a heavy, hard-boned face and tousled black hair, a half-blooded Creek from Enid who, for occult reasons of his own, had made up his mind to die of pneumonia. (p. 43)

"Again, this is Chief Myers. He was every one of those words except he was a half-blooded Comanche from Apache, Oklahoma."

COLONEL MOODUS

Colonel Moodus was General Dreedle's son-in-law, and General Dreedle, at the insistence of his wife and against his own better judgment, had taken him into the military business.

General Dreedle gazed at Colonel Moodus with level hatred. He detested the very sight of his son-in-law, who was his aide and therefore in constant attendance upon him. He had opposed his daughter's marriage to Colonel Moodus because he disliked attending weddings. . . (p. 36)

"War is hell," he declared frequently, drunk or sober, and he really meant it, although that did not prevent him from making a good living out of it or from taking his son-in-law into the business with him, even though the two bickered constantly. (p. 212)

About Lt. Robert D. Knapp, Jr.:

"I was back at 57th Wing Headquarters on schedule. I rounded up a crew, which now included Bob Jr. (General Robert D. Knapp's son, Lt. Robert D. Knapp, Jr.) —new 2nd Lieutenant—as my Aide,"[18] confirmed Bob Knapp.

DOUGLAS ORR

"You [Yossarian] really ought to fly with me, you know. I'm a pretty good pilot, and I'd take care of you."

"Will you fly with me?"

Yossarian laughed and shook his head. "You'll only get knocked down into the water again." . . .

Orr did get knocked down into the water again when the rumored mission to Bologna was flown, and he landed his single-engine plane with a smashing jar on the choppy, wind-swept waves tossing and falling below the warlike black thunderclouds mobilizing overhead. He was late getting out of the plane and ended up alone in a raft that began drifting away from the men in the other raft and was out of sight by the time the Air-Sea Rescue launch came plowing up through the wind and splattering raindrops to take them aboard. Night was already falling by the time they were returned to the squadron. There was no word of Orr. . . *(p. 310)*

"Orr?" cried Yossarian.

Washed ashore in Sweden after so many weeks at sea! It's a miracle. "Washed ashore, hell!" Yossarian declared, jumping all about also and roaring in laughing exultation at the walls, the ceiling, the chaplain and Major Danby. "He didn't wash ashore in Sweden. He rowed there! He rowed there, chaplain, he rowed there."

About Douglas Orr, Bombardier/Navigator:

Douglas Orr was the only character in *Catch-22* who retained his own name. While returning from a mission, bombardier/navigator Douglas Orr's aircraft was downed over the Mediterranean. Orr and the rest of the crew barely had time to throw out the rubber raft, with its little blue oars, and grab their jungle kits. They began paddling towards Tunisia and after twelve hours, guided by Orr, the raft reached land behind enemy lines. *The Stars and Stripes* newspaper reported how these shipwrecked fliers, still led by Orr, took cover in the tall brush. At one point they encountered a herd of goats, which they milked and then slept with for warmth. They walked for four days, wore out their shoes on the rugged terrain, and eventually had the good fortune of making contact with an advance British patrol.[19]

In Heller's autobiography, he reveals that he based Orr's character also on a pilot named Edward Ritter who shared Heller's tent. Again, two personalities have merged. Ritter was a "tireless handyman", building the elaborate fireplace with a mantel in their six-man tent, assembling their gasoline stove, and creating a washstand out of a bomb rack and a flak helmet. Just like Orr, Ritter also had a penchant for ditching in the water and crash-landing safely on land without losing a single crewman:

"Remarkably," says Heller, "through all his unlucky series of mishaps the pilot Ritter remained imperviously phlegmatic, demonstrating no symptoms of fear or growing nervousness, even blushing with a chuckle and a smile whenever I gagged around him as a jinx, and it was on these qualities of his, his patient genius for building and fixing things and these recurring close calls in aerial combat, only on these, that I fashioned the character of Orr in *Catch-22*."[20]

KID SAMPSON

Even people who were not there remembered vividly exactly what happened next. (p. 331)

. . . prepared for any morbid shock but the shock McWatt gave him [Yossarian] one day with the plane that came blasting along the shore line …

George: "Now this is what I think Heller did: He took two actual incidents - one, the only plane on a non-combatant flight to crash into a mountain and two, a mid-air collision that cut most of the tail off of one of the planes. The only person killed in that plane was the tail gunner, who was cut right out of the tail and never heard of again. Heller tied these two incidents in with an actual name, pilot Bill Simpson. In *Catch-22* there is a character named Kid Sampson described as blond-pale-scrawny, who was cut in half by McWatt's low flying plane that McWatt crashes into a mountain to pay for his miscalculation. He combined that incident with a mission that saw pilot"...

"Dobbs who 'zigged' when he should have 'zagged', skidded his plane into the plane alongside, and chewed off its tail. His wing broke off at the base, and his plane dropped like a rock and was almost out of sight in an instant. There was no fire, no smoke, not the slightest untoward noise. . . . It was over in a matter of seconds." (pp. 368-369)

George: Heller may have drawn on Simpson, who piloted the plane that cut the tail section off of the other plane, and who was also very young, blond, and scrawny, and made him pay for slicing the tail gunner out of the plane by having him crash into the mountain.[23]

MAJOR MAJOR MAJOR MAJOR

Captain Black ... maintained that Major Major really WAS Henry Fonda but was too chickenshit to admit it. (p.88)

About Major Major and Capt. Randall C. Cassada:

George: He is Randall C. Cassada and he was a captain and squadron commander when I arrived in the squadron.

Heller says, "Major Major Major Major (the squadron commander) was long and bony - looked a bit like Henry Fonda in distress" [22], a description that fits Cassada. And like Major Major Major Major he also played basketball once in a while and was withdrawn by nature. Didn't talk much. In the Group there was also another officer named Major Major, upon whose name Heller drew. The squadron was, in my opinion, the best in the Group. Most of the people who ended up in key jobs came from the 488th. That movement of key people alone justified the fact that the 488th was the top squadron in the top Group in the top Wing in World War II. So Cassada must have been doing something right."[21]

THE DEAD MAN IN YOSSARIAN'S TENT

Originally Yossarian bunked with Orr and "The Dead Man" who died before he even got there.

Mudd was the unknown soldier who had never had a chance, or that was the only thing anyone ever did know about all the unknown soldiers—they never had a chance. They had to be dead. And this dead one was really unknown, even though his belongings still lay in a tumble on the cot in Yossarian's tent. *(p. 107)*

About The Dead Man In The Tent:

The sad story of "The Dead Man in Yossarian's Tent" was based in truth. A soldier had reported in, had been sent on a mission, and had been shot down and killed before his orders had even arrived. He - actually, his belongings- remained in the tent until the red tape cleared "him" for return to the States.

Heller, like Yossarian, also had a vacant cot in his tent until pilot Ritter claimed it. This cot had belonged previously to a bombardier named Pinkard, who, on a mission over Ferrara, was shot down and killed.

TARANTO

. . . the alert sounded suddenly at dawn the next day and the men were rushed into the trucks before a decent breakfast could be prepared, and they were driven at top speed to the briefing room and then out to the airfield, where the clitterclattering fuel trucks were still pumping gasoline into the tanks of the planes and the scampering crews of armorers were toiling as swiftly as they could at hoisting the thousand-pound demolition bombs into the bomb bays. Everybody was running, and engines were turned on and warmed up as soon as the fuel trucks had finished.

Intelligence had reported that a disabled Italian cruiser in dry-dock at La Spezia would be towed by the Germans that same morning to a channel at the entrance of the harbor and scuttled there to deprive the Allied armies of deep-water port facilities when they captured the city. For once, a military intelligence report proved accurate. The long vessel was halfway across the harbor when they flew in from the west, and they broke it apart with direct hits from every flight that filled them all with waves of enormously satisfying group pride until they found themselves engulfed in great barrages of flak that rose from guns in every bend of the huge horseshoe of mountainous land below. Even Havermeyer resorted to the wildest evasive action he could command when he saw what a vast distance he had still to travel to escape, and Dobbs, at the pilot's controls in his formation, zigged when he should have zagged, skidded his plane into the plane alongside, and chewed off its tail. His wing broke off at the base, and his plane dropped like a rock and was almost out of sight in an instant. *(pp. 368-369)*

The following three photographs and two documents chronicle the astounding success of the rise and fall of the Italian cruiser, *Taranto*.

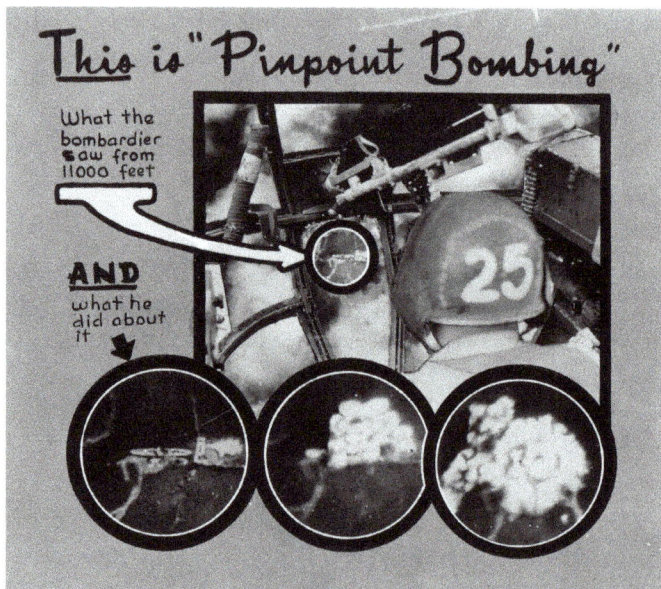

What the bombardier saw from 11,000 feet.

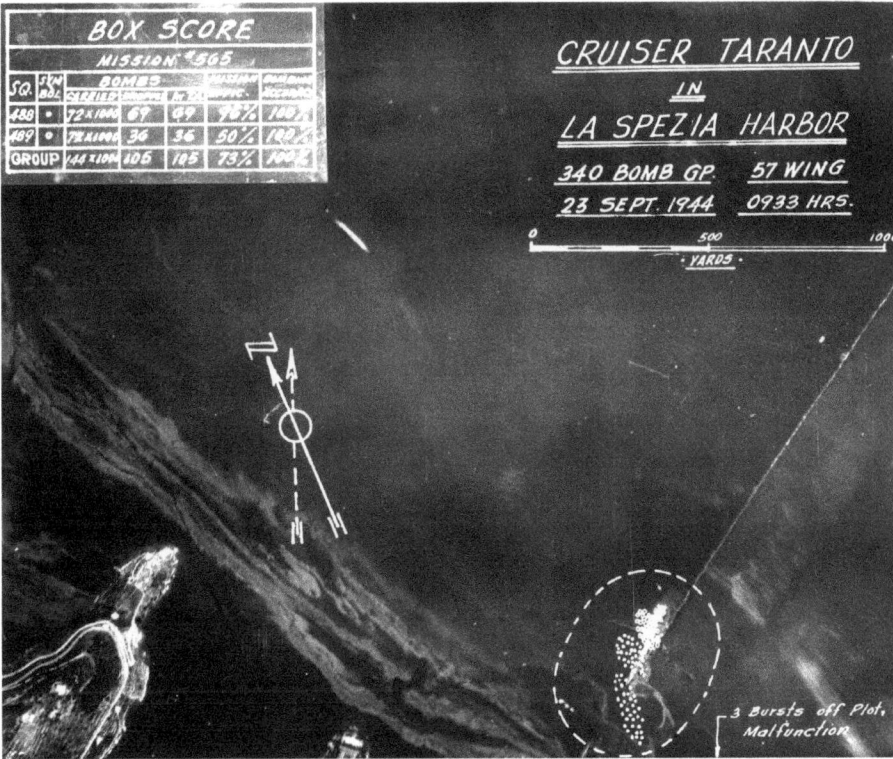

The official photo of the bombing as it occurred.

Photo of submerged and useless *Taranto*.

C. O.

PHOTO INTERPRETATION REPORT
340th Bombardment Group
Mission #565   23 September 1944

    The target was the light cruiser Taranto in La Spezia harbor.
The 488th's first pattern was very compact and started on the cruiser.
Several direct hits amidships were both observed and photographed.
The second pattern, 488th, crossed the bow, and the third, 489th,
crossed the stern. Hits were probable, but could not be seen. The
fourth box, 489th, returned its bombs because the cruiser was entirely
obscured by smoke. *Recce photo shows foredeck awash*

Mission #566   23 September 1944.

    When we bombed the Cittadella R.R. bridge at G-238736 on 31
August, our photos revealed that a rail bypass or diversion was
being consturucted across the riverbed. That diversion, now completed,
was the target for the 487th squadron. Only the first box bombed
here, releasing by radio. The pattern was just short of the east
end of the diversion.

    The second box bombed the alternate target, which was the Padus
East railroad bridge at G-395509. The pattern bracketed the track
650 yards east of the bridge.

Mission #567   23 September 1944.

    This mission's target was the Piazzola R.R. bridge at G-290834.
Two boxes of the 486th squadron attacked it. The first pattern
crossed the west approach and the second crossed the tracks 350
yards east of the bridge. Both probably cratered the track.

                                   FRED C. EGGERS,
                                   Capt., Air Corps,
                                   Photo Interpreter.

Official interpretation of the technical details of the direct hits.

R E S T R I C T E D

HEADQUARTERS TWELFTH AIR FORCE
APO 650

GENERAL ORDERS)   27 December, 1944
                :
NUMBER   281)

Under the provision of Circular 333, War Department, 143, and circular 89, North African Theater of Operations, 10 July 1944, the 340th Bombardment Group (M) is cited for outstanding performance of duty in action against the enemy in the Mediterranean Theater of Operations on 23 September 1944

Culminating a long and unbroken series of flawlessly executed bombing attacks on pinpoint and area targets, the 340th Bombardment group (M), in sinking the enemy light cruiser Taranto on 23 September 1944, distinguished itself by such extraordinary heroism and professional competence in the face of vigorous enemy opposition as to set itself above and apart from other units participating in similar operations. On 22 September 1944, when aerial reconnaissance disclosed an enemy plan to scuttle the Taranto at the entrance of La Spezia Harbor in Italy, the 340th Bombardment Group was ordered to destroy this warship with all speed before it could be moved into position. Acting swiftly and with utmost thoroughness, the Group's operations, intelligence and maintenance personnel skilfully planned the attack, briefed the crews, and readied their planes for the assault. At 0300 hours on 23 September, 24 B-25's of the 340th Bombardment Group took off from bases in Corsica for the heavily defended La Spezia area. Despite heavy anti-aircraft barrage from the ring of powerful enemy batteries which encircles the harbor, gallant pilots, displaying outstanding courage and flying ability, resolutely held their aircraft in tight formation throughout the attack. Highly trained bombardiers, undeterred by the hostile fire, expertly synchronized their instruments and released their thousand-pound bombs with unerring precision, scoring numerous direct hits on the target. Compact patterns from the first three flights covered the bow and stern of the cruiser with devastating effect, capsizing it before the last flight could release its bombs. This outstanding achievement, made possible by unsurpassed teamwork which combined exceptional planning with indomitable courage, flying skill and precision bombing, completely frustrated the enemy in his attempt to block the entrance of this strategic harbor and naval base. The heroism and extraordinary professional skill displayed by the 340th Bombardment Group in this action reflect highest credit upon themselves and the Military Service of the United States.

By command of Major General Cannon

CHARLES T. MYERS
Brigadier General, USA,
Chief of Staff

OFFICIAL
/s/ William W. Dick
    William W. Dick
    Colonel, AGD
    Adjutant General

/s/ THOMAS B. MYER
    Captain, Air Corps.

Official citation for the 340th Bomb Group's outstanding performance.

PART III

THE 340TH BOMB GROUP SPEAKS:
TWELVE TRUE TALES

Rear section of
aircraft.
" Bailing out"
procedures

The hourglass sands are sifting . . .

The effects of World War II were incalculable. Countless lives, if not ended, were altered. character and resolve of nations had been tested. Our world was changed.

Films, documentaries, and interviews chronicled the events, and experts attempted, f every possible perspective, to interpret them. But the truths come most accurately, clearly insightfully from those who were there, those who saw with their own eyes and heard with t own ears the experience of war. These are the ones who left their mark and upon whom m were left. They are now history's treasures. The hourglass sands are sifting, but it is possible to hear some of these stories firsthand.

Joseph Heller, once he was years removed from the war's drama, treated his experie with humor—albeit of the dark variety—in his creation of *Catch-22*. Others from his 3 Bomb Group also penned their remembrances. Some are related with humor as well, some pathos, some from a need to document and preserve, but all with the uniqueness of eyewitness. They have their own stories to tell, their memories that need to be shared. And t perspective is like no other.

Heller, with his *Catch-22*, has earned his spot in the sun. Now we can pay close attentio a few of his war mates, for herein lies The True Story.

Since the 340th was momentarily brought to its knees by Mt. Vesuvius, there have bee many versions of those hours as there are men in the group. It was historic and it was perso Earlier in this book there have been accounts or references to the event, but the following co from a different perspective.

CHAPTER 7

TALE #1

PAUL GALE SPEAKS: 489TH NAVIGATOR
VESUVIUS SPEAKS — WE LISTEN

It was mid-March, 1944. We were quartered in mud huts in the nondescript village of Pogio Marino at the base of Mr. Vesuvius. That volcano had been announcing its instability with spasmodic, fiery showers for several days.

I was scheduled to leave for R&R in Cairo on the 9th. I packed my things in a barracks bag, threw it in a corner, and asked the fellows to toss it on a truck as and when "that thing blows its top." They threw a couple of shoes at me and I left.

Our designated transportation was a used-up, battle-weary, flak-ridden B-25. Complaints to our CO, Major Kaufmann, produced a "That's it, guys, it's all we can spare—you don't have to go." We went.

There were three of us: Roberts, a copilot whose name I have forgotten, and me, the stalwart, ever-dependable navigator.

We arrived in Cairo on the 20th, parked the plane at Payne Field, and rode the bus to town. The fresh steak and huge tomatoes were a treat —something that was not available at the officer's mess at the field. We asked why. "Probably because they don't like the way we fertilize them." And we thought the "honey bucket" brigades just quaint, colorful background. We stopped eating tomatoes.

On the 28th a telegram came ordering us back. Vesuvius had blown, and —get this—we had the only serviceable aircraft.

We headed back, but not before loading the bomb bay with Egyptian beer, wire swing bales securing faulty ceramic tops to old wine bottles.

Desert flying is hot, very hot. So we headed up to cooler, comfortable air. It was not only more comfortable, it was downright aromatic. The bomb bay was foaming, delightful, delicious, malt-laden foam.

We went down. Down was a hot and thirsty place; the beer was cool and going flat. We drank the beer, we sweated and went up. The beer foamed, we went down, sweated and drank the beer. And so it went. The profile of our flight was a continuous series of erratic ups and downs. We were dedicated and doing our level best (no pun intended) to keep that beer from going flat.

In due time we arrived at the designated ash-covered field. A strip had been cleared for us and a VIP group of brass was anxiously awaiting our arrival. After all, we did have the only operational aircraft in the group.

Not ones to flaunt tradition, we buzzed the field. Blowing volcanic ash over one and all, we landed in a series of staccato bounces.

We tumbled out of the plane as our not-too-friendly, red-faced CO drove up. We greeted him with all the dignity at our disposal. "Here's your _____ing plane, Major, sir."

From a personal diary: ". . . we found almost complete devastation. Tents were torn to ribbons and 88 airplanes were a total loss. Eighty-eight B-25 Mitchells — $25,000,000 worth of aircraft. How Jerry gloated. Axis Sally[2] dedicated her program one evening to the survivors of the 340th Bomb Group. Actually a sprained wrist and a few minor cuts were the only casualties. The following night she cracked, 'We got the Colonel. Vesuvius got the rest.' She explained how the 340th was no longer operational. How wrong she was. Within a week the 340th was again bombing Jerry in Northern Italy."

CHAPTER 8

TALE #2

WARD LAITEN SPEAKS: 487TH CREW CHIEF
I AM 7-K
THE EARLY BIRD

I was a B-25-C S/C 42-32278, assigned to the 340th Bomb Group, 487th Squadron, on a cold, snowy, icy field at Battle Creek, Michigan (Kelly Field) on February 12, 1943. Seven men came up to me and walked around looking me over. They looked inside my wheel wells and bomb bay, then opened up the hatches and got in, looking me all over inside. I found out the names of these men a few days later. They were to be my crew. The crew included: Marshall E. Lambert, pilot; Harley H. Anderson, copilot; Robert Sather, navigator-bombardier; Theodore Handzel, engineer; Maurice M. Schwartz, radio-gunner; Woodrow W. Peterson, armourer-gunner; and Ward Laiten, crew chief.

The next few days found the men all over me, cleaning guns and loading supplies. The crew chief's inspection found that my right outer wheel bearing was missing. This had to be installed. As the wheel bearing was being replaced one of the men noticed that I was painted pink, which meant that we were all headed for the Desert War Zone.

It was below zero the morning that we were to leave Michigan. I hadn't been test hopped. My crew climbed aboard and tried to start my engines. From 7:00 a.m. until 11:00 a.m. the men took turns cranking my engines until they finally turned over. We were off for West Palm Beach, Florida (Morrison Field); temperature, 65 degrees. My pilot set me down on the ground and while some of the others loaded me (even replacing my guns), the crew chief changed my gaskets in order to repair leaks.

One sunny morning we took off for Puerto Rico, but the weather turned bad and we flew through, over, and under thunderstorms before we made the island. A few more oil leaks were discovered, and when my crew chief removed my bottom rocker box covers he found the rocker arms had been so hot they turned blue. While we waited for parts Andy had the crew's names painted on me and I was named *The Early Bird*. The new gaskets finally arrived and the crew chief worked all night to install them.

The next morning the crew headed me for South America, landing in

Trinidad for fuel and changing one of my voltage regulators. Then on to Atkinson Field, Georgetown, British Guiana. When we were getting ready for take-off the next morning, I blew my right starter. The crew had to change a new one.

Finally, on 1 March 1943 we left for Belem, Brazil. Heavy rains kept us from landing until late in the afternoon. After a couple of days' rest we took off for Natal, Brazil. About an hour out my left engine began to backfire, so we turned back to Belem. The crew changed my spark plugs and discovered someone had put a lot of water in my tanks. They drained my sump and wondered who had done this to me. We finally made it to Natal, where I had a fifty-hour inspection.

On 8 March we took off to cross the BIG pond, leaving a lot of my equipment behind as well as my crew chief and one gunner. We had to cut down on the weight, for it was a long trip to the Gold Coast of Africa. I waited there until my wing racks were installed. The gunner installed .30 cal. guns in my tail cone, hooked to the top turret that watched my tail.

On 20 March I took off for my first flight over Africa, flying over sandstorms and landing at Maiduquria, Nigeria. We gassed up and left for Khartoum, landing there with one of my engines running very rough. The crew changed #5 cylinder plugs in my right engine and it smoothed out. The day after, the crew did a twenty-five-hour inspection.

We landed at Wadi Halfi on the Nile River, gassed up, and installed new plugs in the #5 cylinder. A few miles after taking off again, the engine began acting up, forcing us to return to Wadi Halfi. After a few days' wait, the crew got a set of rings and gaskets to replace the old ones.

The next day we headed for Heliopolis, Egypt, but before we reached our destination the engines got rough again. The crew found that I was burning oil very badly; all my plugs needed to be changed again. The crew unloaded all the weight possible and off I went to join my sister ships at Al Kabrit on the Red Sea. I smoked like a steam engine and the flying got rough again. The right engine and the oil in the left engine were changed between sandstorms, and then I flew on the Costal Bisito (Tripoli) near the war zone. That night the crew chief and engineer began sleeping under my wings at night in the sand and heat. They could not touch me during the day, as I would get so hot from the sun.

On 2 May, I took the crew from the hellhole to Sfax and rejoined the squadron and prepared to do my duty.

10 May 1943! I, 7-K, *The Early Bird*, flew my first mission carrying eight 250-pound bombs. I carried these English bombs over the island of Pantelleria. My second mission didn't come until 30 May. I had become

known as old stand-by, having my bomb load changed many times a day.

I moved to Hergla on 3 June, making two trips to haul equipment. Some Brass spotted my tail guns and because he hadn't approved them, had them removed. They had worked great in the past to keep the enemy off my tail.

I began to fly missions over Sicily on 4 July and took my first flak on 29 August. On the 29th the crew and I moved to Catania, Sicily, and the next day I flew over Italy for the first time. We continued to fly missions as assigned to us. We learned, on 9 September, that Andy Anderson, who had named me *The Early Bird*, was found dead in the water after a mission over Naples. My crew felt very bad about this news.

We endured the mud, rain, and all the other ills of a combat operation, moving from base to base as required. So, on 3 January I moved the crew and all of the equipment again. Everything had to be moved when I moved. The crew even loaded some of the equipment on my wing racks. What a load!

I was to land at Pompeii, Italy, but first we circled Mt. Vesuvius. Our airfield was just east of the mountain. Up to this date I had dropped 223,260 pounds of bombs on our enemy. I saw many of my sister ships go down as I flew over the Anzio beachhead.

Because my crew chief didn't like oil on my engines, cowlings, or landing gear, the ground crew always changed my gaskets every time I returned from a mission. One of them was always cleaning my guns, keeping them oiled. I learned that my crew chief left to become a crew chief on one of my sisters. I'll miss Ted Handsel; he was a good assistant.

They installed a Norden bombsight on March 1944 and I became a lead airplane. This meant flying many times a day as a practice ship, getting the bombardiers used to the new bombsight. I also flew my share of missions.

Then came disaster! On 22 March 1944, at 2:30 a.m. Mt. Vesuvius blew up. Ashes dropped on me, putting small holes in my fabric controls. At 4:30 a.m. another eruption occurred. More ashes put bigger holes in the fabric. About 6:30 a.m. red-hot cinders about the size of a crewman's fist dropped on me.

Orders were given to remove all my controls, but before the crew could get all of the cotter pins out of my bolts, the mountain thundered and red-hot ashes as big as a man's head flew through the air. Everything was covered with two feet of ash. It was the end of all of my sisters on the field. We had holes in our wings; our windows and windshields were all broken.

I, 7-K *Early Bird*, had flown 90 missions, some 444 hours, with only one flak hole. This day was a sad day for the ground crew who had taken care of me. The ground crew moved off to the 321st field near Paestum. My crew chief and five other men stayed behind to dig equipment out of the ashes. My crew chief cleaned the ashes off my wings and took my clock out of the panel, then he left me for others to take care of.

It was the end for me.[1]

CHAPTER 9

TALE #3

HERB BARCLAY SPEAKS: 488TH RADIO GUNNER
CHAFF OVER THE TARGET

On December 30, 1944, our B-25, from the 488th Bomb Squadron, was assigned a mission completely new to our crew—to fly the number two position of a "chaff" element, part of a 340th Bomb Group mission to bomb the railroad bridge at Calliano, Italy. None of our crew had ever handled chaff before.

Chaff was a long strip of tinfoil on paper backing, banded and bundled in sizable clutches, and packed in cartons. The idea was to dump the stuff into the slipstream where it would scatter and show up on the enemy's radar as a blip, deceiving the anti-aircraft guns. The sheet metal boys had rigged up ingenious chutes in the waist escape hatches of two of the chaff ships —the other two, not ours.

Our crew for this mission included Bob Gilliam, who I think was flying his first mission as first pilot. The copilot was possibly Clayt Chambless. Al Silverman was bombardier, whose task this mission would be to relay instructions from the cockpit to tail gunner Buck Norris and myself, the radio gunner, who would be actually handling the chaff, and engineer Wil Proulx.

As we came up on the IP, Al passed us the word to get ready. Buck and I positioned ourselves. I stood over the rear hatch, facing forward. Buck, standing in front of me, would hand me bundles of chaff from the cartons which we'd ripped the tops from so as to be able to throw at the four-second intervals they'd instructed us in the briefing. Later we were to regret removing those carton tops.

Al said, "Let her go," and I threw the first bundle at the escape hatch. That's as far as the chaff got. The slipstream kept the stuff from ever leaving the plane and blasted it back into the tail section. I looked at it in amazement, but Buck was already shoving the next bundle at me. I threw that one a little harder, but it still wouldn't leave the plane and ended up in the tail, which was starting to look like a Christmas tree.

I'd just about figured how to jam my hands out the hatch, where the slipstream would pull the chaff away, when the flak started to show up. A big black .88 exploded right where I was going to deposit the next bundle. They had our range perfectly.

We managed to get a little more out when, all of a sudden, it seemed we were going straight down. Buck and I and the cartons of chaff were on the ceiling. Next thing, we came out of the dive and climbed straight up. Buck and I and the chaff were thrown to the floor and held there, by I don't know how many Gs. By the time we went into our next dive, the chaff all came floating out of the cartons and filled the ship.

My mike and earphone cords had become unplugged and I was out of contact with the cockpit. I thought sure we were hit and were going in. I looked for my chute, but couldn't find it. It must have been floating somewhere in that sea of tinfoil.

The flak stopped, and we finally leveled off. For a while I lay in the pile of tinfoil, too exhausted to move. Every inch of the plane was covered with foil, which had even gone up over the bomb bay and into the cockpit.

Buck and I spent the next two hours trying to clean up the ship. We were afraid to face the crew chief with his plane looking like a Christmas tree.

Back on the coffee-and-doughnut line, the crews of the other chaff ships laughed and asked what the hell we'd been doing. Evasive action, we answered, matter-of-factly. Then they told us what had happened from their point of view.

When we'd taken our first dive, the guns on the ground must have picked us up, because we were followed by bursting .88s all through our up-and-down evasive action. The other crews hadn't drawn another shot and had calmly gotten rid of their chaff.

As it turned out, we hadn't been able to handle the chaff, but we did take the flak, which allowed the other crews to dispense their chaff and help the bomber crews after all.[1]

TALE #4

FRANK B. DEAN SPEAKS: 380TH GROUND CREW
WE LOSE THE QUESTION MARK

The ritual of watching our B-25s return from the mission had started long before the twenty-two-plane formation had appeared as tiny black dots on the horizon. Anxious eyes searched the formation for gaps in their ranks that indicated missing aircraft.

Ray Conrad, Leon George, and I knew our plane was missing long before anyone else. From the moment we spotted empty space on Capt. Cometh's right wing, we knew that it was Lt. McCormick and his crew. With this knowledge came that old familiar feeling of fingers of fear clawing at your guts and the sad, sick, empty, hollow, half-crying, throat-choking, breath-squeezing anxiety that mingled with the "Lord, don't let it be" desire to make the brain refuse to believe what the eyes were seeing.

TARGET SARDINIA

There were two gaps in the formation but we were only concerned with the one that appeared in the twelve-plane section of our 380th Squadron, where earlier the *Question Mark* had snuggled into position with Lt. Cromartie to form Capt. Cometh's "A" flight. The target had been the enemy airdrome at Villacidro, Sardinia.

The mission of the 310th Bomb Group, on 21 May 1943, had started with the usual preparation. Although it was an evening mission, Conrad, George, and I had risen early, checked the bomb bay where clusters of twenty-pound fragmentation bombs hung from the racks, inspected, pre-flighted, and made our plane ready for flight.

The gunners had arrived, stowed their parachutes, and checked their guns and ammunition supply before sitting down to chat awhile before the rest of the crew arrived. The officers and bombardiers had stayed at Headquarters to be briefed on the target, the approach, altitude, weather, flak gun concentrations, escape routes, and other pertinent mission data.

We had walked through the shakedown inspection with Lt. Robert

McCormick, obtaining information on the target and the mission, while he questioned us about the airplane. Lt. Norman Toenjes had settled into the copilot's seat to make his required checks. T/Sgt. Dave Richardson had crawled through the metal tunnel into the bombardier's position in the Plexiglas nose of the aircraft. S/Sgt. Frank Oliver had climbed through the rear hatch to do whatever radio gunners do to test their equipment. S/Sgt. Anthony "Tony" Leanza had followed and wedged himself into the upper turret to make his final gunnery checks. In time for the 3 o'clock takeoff, the engines had whined, coughed their usual blue smoke, then settled into the familiar sound that Wright "Cyclone" engines, with short exhaust stacks, always make.

There had been the final hand waves from us to the combat crew that expressed our unspoken concern for the men we considered ours. They had returned them with half-salutes, completing the other half of the circle that bound them, the plane, and us, into one. We had watched as the airplanes had rolled down the runway, lifted their noses, and climbed into groups of threes, then sixes. They had circled the field until all twenty-four planes were in formation, then headed northeast toward the island of Sardinia, where the Italians had reinforced their *Regia Aeronauticia* with additional MC 202s as well as Messerschmitt Bf 109Gs. It was hoped that they could be caught on the ground.

ESCORT P-38S

We of the ground crew had watched as the formation turned into black dots on the horizon, then vanished. We would begin our vigil of searching the sky long before their scheduled return around 6:00 p.m.

The bombers with their P-38s for escort had crossed land and flown low over the blue waters of the Mediterranean, until the southwest coast of Sardinia became a dark smudge on the horizon. The bombers had tightened their formation and climbed up to their bombing altitude of 9,000 feet. The escorting, twin-tailed fighters had pulled into position of defense on the perimeter of the formation.

Little attention was paid to geography as the American planes droned across the shoreline into enemy-held territory. Gunners and fighter pilots anxiously scanned the skies for fighters wearing black crosses or three fasces in a circle. Bomber pilots divided their attention between staying in close formation and searching the forward segment of the sky. The eyes of copilots moved over the land and sky, watching for danger.

Far below, the alert had been sounded. The enemy would not be caught napping. Pilots raced for the cockpits of their mottled, single-engine fighters, while anti-aircraft gunners feverishly prepared for the onslaught.

The target came in view. Crewmen spotted the river, the hills, the runway, and taxi strips. Ten or twelve large aircraft were parked in the northeast corner of the field. Across the road, many large bombers and transport planes were scattered about

ACCURATE FLAK

Intense, heavy, accurate bursts from anti-aircraft guns blossomed among the bomber formation, as Axis fighters locked in aerial combat with American P-38s.

Smoky flak bursts, aerial dog fights, red streams of tracers, hammering machine guns, excited voices calling out enemy fighter positions over airplane intercommunication systems, open bomb bay doors, clusters of fragmentation bombs pouring from the bellies of olive-drab colored bombers, white knuckled hands beneath soft leather gloves turning control wheels or moving joy sticks, steel projectiles piercing aluminum skin or human flesh, smoke, fear, prayers, curses, dry lips, pounding hearts, and crescendos of sound were woven into the fabric of the bomb run.

Nine thousand feet below, a sweating soldier in a baggy grey uniform slammed a steel projectile into the smoking breech of an anti-aircraft cannon as the formation became locked on course and altitude as the bomb run started. Gunners cranked wheels as the sights centered on a miniature bomber flying on the right wing of a three-plane flight.

Burning gases hurled the twenty-pound shell skyward, toward its target. The point of the speeding projectile pierced the thin, aluminum covering and the rubber wall of the wing tank of the twin-tailed bomber with a question mark painted on its nose. The fuse triggered a blast that flashed fire into a fuel cell filled with fumes and 100-octane aviation gasoline.

One moment it had been a B-25 bomber on a routine bombing run. The next instant it peeled off to the right, dropping away from the formation with red tongues of fire gnawing at the fuselage and flames billowing and flowing in a fiery, sixty-foot tail. The *Question Mark* vanished in a red-cored, roiling, black-bordered ball of red-hot explosion that devoured men and machine and flung pieces of flaming debris that trailed incendiary smoke, as they arced downward toward the Sardinian countryside. These streaks of fire that crossed the heavens like runaway comets, marked the funeral pyre of the men we had known.

Friends anxiously searched the flaming wreckage for signs of parachutes. Lips mouthed curses, prayers, and words of encouragement of,

"Come on, come on." They counted four parachutes, but Lt. Beachum's plane was plummeting to the ground, and it was impossible to determine who was hanging safely under the open canopies.

S/Sgt. Paul Kessler, riding in the bombardier's compartment of Lt. Sutt's plane, aimed his camera at a flaming, galling gob of fire and froze a brief moment of history in the emulsion of a film.

There was little time to check survivors, as flak bursts and enemy fighters were harassing the formation. Three sharp-nosed fighters bored in for attack on the bombers. Two were driven away and one was shot down.

The remainder of the formation droned toward the sea and their base at Berteaux, while enemy fighters continued to try to burst through the P-38s riding shotgun to get at the B-25s as they pressed on toward the coast.

The airplanes flew across the sea toward home. The only obvious evidence of sudden death in the May afternoon were the two gaps in the formation where the planes piloted by Lt. McCormick and Lt. Beachum had once been.[1]

CHAPTER 11

TALE #5

R. M. JOHNSTON SPEAKS: 488TH PILOT
SHORT SNORTER

From the left seat of the war-weary B-25C the Limey brass nodded out the open window to a "little" dirt strip off to port. Nestled on the lee side of a hill, from 500 feet it looked more like a cart path through an olive grove than a fighter strip. Letting go of the wheel, he pointed and ordered, "Land at that Spitfire strip, old chap."

I almost asked "What!???....*there?*" as my common sense roared, "You gotta be out of your bloody 'limey' mind." That dirt path looked like a tight fit for an L-5 as I hurried to drop the flaps and roll into a dragging, slow approach while thinking, "How the hell did I get into this?"

An hour before I'd been lounging in the shade outside squadron ops, back at Comiso. We had been waiting since morning to scramble on a close support mission, but the target kept changing. I had been half asleep when our operations officer Homer Howard called, "Hey, RM! There's some limey brass on his way, and he wants to be taken up toward the bomb line somewhere."

I didn't care who the limey was; it got me off stand-by and sure as hell beat flak and fighters. As I approached my plane, or 8A, an open British staff car with pennants flying and a roar of escort motorcycles slid to a stop. A top-ranking limey officer with shoulders full of pins and ribbons to spare bounced out of the car, walked toward my feeble highball, and scrambled up the ladder into the aircraft. Since I was to be his chauffeur, I hurried to follow. Before I could get up the ladder, he had tossed his briefcase on the deck, jumped into the left seat, hit the starter switch for the left engine and asked, "I say, old chap, I haven't flown a Mitchell for years . . . do you mind?"

Now I ask myself, "How th'hell does a 2nd Lt. say no to an 'ol' boy' wearing all that brass?" Before I could even get myself settled in the right seat, he had both engines running, and kicked off the brakes as we started to roll. This left Sgt. Mario Vuotto, my crew chief, racing underneath, trying to get up the hatch when my pilot asked, "I say, which way to the strip?" I managed a curt, "Turn left, sir."

By then Mario, panting like a marathon runner, tapped my shoulder with a thumbs-up. At the end of the runway the "ol' boy" didn't look for traffic, slow down, or check the mags. On the run he rammed the throttles to the firewall and let her go. Before I could get booster pumps, manifold pressure, etc. cleaned up, he horsed 8A into the wild blue with no flaps, and at minimum speed in the hot August air. Needless to say, we weren't climbing too well! We were a yard behind the power curve with the hilltop town of Comiso directly in front of us, and way above eye level. Now, I'm not Catholic, but silently reciting a Hail Mary, I sneaked on some flaps and tried to reset my pucker factor below the red line. We roared right up Main Street, level with the second floor while "Pizon" waved wildly from their balconies.

At 500 feet, we wandered aimlessly along the bomb line looking at the sights like bloody tourists. I didn't mind the tour, only the ground fire we were drawing. After twenty minutes of gawking, we flew a few sweeping circles and he pointed out that short—I mean short little Spitfire strip in the olive grove. Suddenly with his seat back, a half-mile short of touchdown, he turned the controls over to me, and it was "all mine!" He was ready to get out.

I slammed her down, nose high, full flaps and full power just short of the markers, and while braking hard, slid round a ground loop at the end of the strip in a boiling cloud of dust. Before I could even shut her down, the ol' boy was in the car and motioning for me to hurry. Leaving Mario to clean up, I rushed for his waiting car. Like a scalded cat, we took off to race through dusty back roads, roaring through a couple British roadblocks, then through two dozen MPs into an olive grove surrounded by a stonewall, and crowded with rank.

In front of a familiar-looking caravan was a long table with white linen and full silver service. I recognized Monty standing aloof amid a dozen white-coated waiters serving drinks. I knew the drill. Standing aside with a drink, I watched my ol' boy being welcomed like a long lost son. With a warm Scotch and water in hand, I sniffed the aroma of good food while trying to blend in with the hired help.

Halfway through my second Scotch, the mess bell rang. To my amazement, the ole boy motioned for me to be seated with all that rank! Me, a "second john" needing a shave, wearing an old 50-mission crush, and in sweaty, dirty suntans with the sleeves rolled up!

Once seated, I recognized General Montgomery sitting at the far end of the table. He had briefed us a few times back in the desert before an operation. But, then I gaped. I couldn't believe it. The head of the table was awash with Monty, Churchill, Sinclair, Wilson . . . then, I heard my ol' boy addressed as Air Marshall Tedder![1] I had only read about him, and wasn't too sure that the Scotch wasn't playing tricks on me. *Me*, at the lunch table with all that brass—like a page in a history book, and I was looking at it.

Head swimming, I listened to the conversation, and it was then I understood what was going on. I was sitting in on a coordinating meeting of the British High Command as they reviewed final plans for the 8th Army and RAF in Sicily. They were discussing the tactical and political plans for Italy. The lunch lasted two hours.

After one more toast, the Air Marshall and I were roaring back to the aircraft. On the speedy drive back, he was pouring through a sheath of papers and scribbling notes on the margins while I was figuring how to get ol' 8A off of 1,000 feet of dirt and over the ridge at the end of the strip. This time, when the Air Marshall bounded into the aircraft he settled in the right seat and continued working on the papers. After I had started the engines, he suddenly looked up as if he had just discovered where he was.

Talking to no one in particular, he directed, "Take me to Malta."

At full power with a running 180-degree turn, I headed down the strip, hauled 8A into the air at the last moment, and we were on our way to Malta.

With no charts, the heading to Malta was by the TAR formula. Remember? You hold up a thumb, sight along it, and guessing, say, "That's About Right." Again, the flight was in complete silence. The Air Marshall, deep in paper work, was probably planning missions for ol' 8A. I didn't care. I was enjoying the memories of that first-class Scotch and chow back in the walled courtyard.

As we made landfall on Malta, I pointed out Spitfires queuing up and, just to be sure, dumped the wheels to show my intent as we went straight in. It was then I remembered the IFF hadn't been turned on all day. I swung 8A around and parked beside a batch of MS's at a staff car beside the sandbagged entrance to operations. The Air Marshall again rushed to get out, but this time the ol' boy waited in front of the aircraft as the props wound down. I thought, "Uh-oh. What now?" as I hurried out.

Walking toward him, I tried to put some words together, but he beat me to it. Holding his right hand out, he stepped forward and said, "Thank you, ol' chap, I'll be on my way now . . . good luck!"

I shook his hand and managed to say, "Thank you, sir!"

I gave him a typical British high ball. Just as he had arrived in the morning, with MP escort, sirens wailing, and pennants snapping on the fender, he disappeared down the road in a cloud of dust.

I filed a clearance, turned on the IFF, and Mario and I headed ol' 8A back to Sicily. With Mario doing the flying, and me feeling a bit relieved, I recounted the lunch story. Then I jokingly added, "Damn! I wish I had asked all them ol' boys to sign my short snorter."[2]

What is a short snorter? It was defined as the bond of friendship amongst the crewmembers or comrades in arms, and it existed typically as paper money signed by two or more men and then separated (torn) so that when all were together again they would still have the money for a drink. It has also been defined in similar terms but consisting of a roll of bills, each from a different man and/or place, and all attached to the next. The Short Snorter has also been described as a sort of drinking club. In order to join, one would have to buy a round of drinks for everyone in the club at the time. If you were in the club and another member slapped you on the back and asked if you were a "Short Snorter," your response would have to be, "You bet your sweet ass I am," and you would have to have the bill to prove it. If you could not prove it you were obligated to buy a round.

Bill Chapman's Short Snorter bills.

TALE #6

BERYL GRAUBAUGH SPEAKS: 340TH PILOT
DOC FIGURES IN ONE OF MY FAVORITE WAR STORIES

O ne night, quite late, Chuck Woodruff came home from the club very drunk. We were about 18 officers living in this bare Italian house about a half block behind 488th Squadron Operations. Inside the entryway we had a pile of 3" steel pipe, which we planned to use for a basketball backboard. I also had an "Indian" military motorcycle, which had been made up out of junk parts.

Woodruff picked up the pipe, dropped it on the tile floor and yelled, "Flak!" Everyone, at least a dozen guys woke up in a panic—and with one single thought: kill Woodruff.

Since I was from the same hometown as Woody, I was always having to save his butt. I finally pacified the situation by reminding them that nobody should hit a drunk, so they only shouted and shoved him around. Then Woodruff threw a roundhouse punch. He didn't hit anyone but he punctured his forearm on the accelerator lever of the motorcycle. No one cared and they all went back to bed. His arm bled and he turned green and puked.

So it was up to me to take him to the dispensary, which was a long walk across a dark, sleeping town. The medic cleaned up his wound and I walked him home and put him in his sack. It was about 3 a.m.

My bombardier/navigator hung out over at the club, and the next day he came back with this story. He said he was sitting at the bar drinking with the doc (Ben Marino) when in came our group commander, Chapman. The doc said, "Bill, I really have to hand it to those boys of yours. A kid came in during the night, a good 8 hours after the mission, with a flak wound in his forearm. We fixed him up, and he's in for a Purple Heart."

I know that Woodruff did not receive a Purple Heart for the fiasco, but he ended up doing some amazing things.[1]

CHAPTER 13

TALE #7
**BEN KANOWSKY SPEAKS: 488TH PILOT
AND MESS OFFICER**
CHARLIE

I remember this guy, Charlie. Charlie was always in the guardhouse.
He wasn't a bad guy but he was always doing these silly little things so
he'd get arrested. Like stealing sugar, nothing really serious. It wasn't until
later, when I was put in charge of censoring mail, that we began to suspect
he was up to something.

Every letter he sent home had money in it—a thousand dollars, fifteen
hundred, not exactly what you would expect from his military pay. The
strange thing was, every time he left the guardhouse he'd be carrying this Air
Force canvas bag. So the colonel told me to follow him. I lost him in
Naples the first time, but the next trip I discovered what he was up to.

While in the guardhouse the prisoners had chores to do, you know,
policing up the area, things like that. They couldn't sit around all the time
doing nothing. Charlie's job was to see that the area was kept clean. He'd
pick up all of the cigarette butts and strip them. His canvas bag was filled
with tobacco and he was selling it in Naples for a very good price. They
were paying him by the pound.

But he wasn't doing anything illegal. And we couldn't legitimately stop
him. Col. Chapman discussed what to do about it. "Don't put him in the
guardhouse,'" I advised the colonel.

Charlie cried when we didn't put him back there. He was knocking off
about 15 thousand a year or better.[1]

TALE #8
HAL LYNCH SPEAKS: 489TH BOMBARDIER
THE BRIDGE AT PARMA

At a pre-mission briefing at 340th headquarters in the autumn of 1944, our CO, W.F. Chapman, called for silence as he prepared to tell us all about the target for the day. He cleared his throat and said, "Our target for today is the railroad bridge near the town of Parma." From my seat near the front of the room I gasped and then exclaimed, "Oh, no!" My reaction had been as involuntary as a sneeze.

Col. Chapman seemed stunned; he turned away from the map and glared at me. "Mr. Lynch, what is your objection to bombing the target at Parma?" he asked.

I replied, "Colonel, I'm sorry I overreacted; it just happens that I love Parmesan cheese and it's a product of Parma."

Most of the men saw the humor and chuckled a bit, but not the colonel.

"Mr. Lynch, if your love for Parmesan cheese is so great, then perhaps you will make a special effort to get your bombs directly on the railroad bridge and not the cheese factory!"

"Yes, sir, I understand. Hit the bridge and not the cheese factory, yes, sir!"

We managed to destroy the railroad bridge at Parma that morning. Not one bomb hit the cheese factory.[1]

Author's note:

After my father's death in 2002, Hal Lynch, in a letter to my mother, included the following: "Bill never selected an 'easy' combat mission during those years we flew over Italy, France, Austria, and Yugoslavia. If it was to be an especially dangerous mission, Bill Chapman could always be found in the lead ship, first over the target, always."

Aug. 21, 1944:
Bombing bridge at
Parma.

Heller and Yohannan
were on this mission

TALE #9
JOHN RAPP SPEAKS: 488TH & 499TH NAVIGATOR
PETRIFIED WITH FEAR

May 12, 1944—Alesan, Corsica, 488th Squadron, 8:30 p.m.

It was a beautiful, warm, balmy spring night, crystal clear, visibility forever, with a brilliant full moon. We were enjoying a rare treat, an outside movie on the hillside. The war seemed thousands of miles away. Suddenly the tranquility of the night was interrupted by a great commotion to the north. Up the coast about 30 miles, towards Bastia, was a British Beaufighter field. We made out the sounds of airplanes and then came the flashes of exploding bombs, each followed by reverberating dull noises. All this lasted for about 15 minutes, and we surmised our night fighter protection had been negated. The movie was called off and we retired to our tent areas under a semi-alert.

I was living in the 488th with Cornelius O'Brien, another captain and navigator, in a two-man tent. We had been checked into the 340th early in March as replacements. Obie, as he had been nicknamed, was a dead ringer then for the present Mark McCumber, a Florida professional golfer. Not long on Corsica we had been warned about digging a foxhole. Our new home put us considerably north of the bomb line and relatively close to access from the Po Valley. Intelligence, however, repeatedly reported there were no German aircraft based in northern Italy. We started digging our foxhole, but not long afterward sort of gave up, thinking, why should we dig up somebody else's island, particularly when the ground was like shale, stratified concrete and full of rocks?

The evening was off to a strange start and my thoughts turned to the next day's mission. It was to be a maximum effort by all of the 12th and 15th in close support around Cassino in an effort to get the bomb line moving again and break the stalemated winter ground war.

We almost made it through the night. Around 4:00 a.m. came the sounds of approaching planes, the telltale unmistakably whirring sounds of German twin-engine aircraft (their props were never in sync). The sirens started wailing and our 40 mm defense anti-aircraft guns began firing.

These were JU-88s, a force of about 40 of them. The first one centered a flare directly over our runway. It hung there like a huge chandelier. The next 20–30 minutes were to become a living hell. They immediately went to work, methodically, pass after pass individually. They got our Group operations first, a small complex area of Butler pre-fab buildings, and then part of our fuel dump. The resultant fires aided by the light of the moon and flare helped them proceed. Then began the main thrust to work over our field and the planes. What came next was sheer terror.

Obie had gotten to the foxhole first, face down, flat prone. I was directly on top of him, face down. Our digging work was perfect as to length and width, but we had quit about four to six inches short in depth. My backside was that much protruding above the hole, out in the open. We became lovers that night, for with every blast I was nudging and furiously pressing into his backside, instinctively trying to get deeper into the hole. They were really working us over and I sneaked several glances toward the field. Brilliant white and yellow explosions. Smoke was rising and blowing everywhere—black, white, gray, yellow and red—whatever had been hit.

The air above us was crazily blowing and gusting in all directions, compressed and propelled by the constant explosions. The ground was trembling and the earth was shaking. The noise and din were the worst of all —deafening and almost unbearable, between the airplanes above and a 40 mm gun fairly close to us on a small knoll, firing like a machine gun with constant explosions. I was never one to be afraid or scared, but that night I was petrified with fear. I thought the world was ending.

Around 4:30 a.m. with bombs all spent, the raid ended. But as a finale,

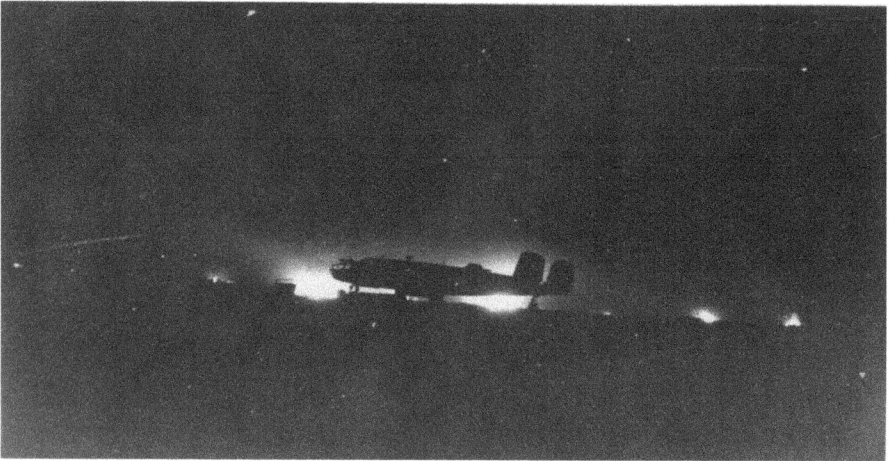

Notation on back of photo: "Taken during air raid 13 May 44"
- Col. Willis F. Chapman

with daylight breaking, some came low with landing lights on and strafed squadron areas. They headed northeast.

So the story goes, we had gotten word over to Italy about our problem and our side got some P-38s off. From high and above they followed the Germans into the Po Valley when they went down to refuel; supposedly our P-38s came down and annihilated them in a veritable turkey shoot. Later we learned where this show of power came from: a roving band, always on the move, for a while on the Russian front, then west to England, back and forth, with one night down to Italy and Corsica.

The next few hours after the raid, I don't recall the casualty count, I vaguely remember we had around 20–25 dead with over 100 wounded.

Our field, a mile-long metal strip for take off and landing, was badly damaged with holes, with the strips bent, torn, and twisted. Surrounding this had been airplanes, some 85–90 B-25s on hard stands, many of these with 500-pound bombs under the bomb bays. Our ordnance had placed them there. Our planes were a shambles. The Germans did a 100 percent effective job on them. Around a couple of sides of the field were rows of trees. They were littered with debris of every shape and pieces of jagged aluminum hung in the branches. There was junk, debris, and litter everywhere.

In about six cases, where bombs were on the ground underneath bomb bays, the planes had completely disappeared, leaving a hole in the ground big enough to start plowing the basement for a new home. My assigned airplane that day, 8A, took a hit in the right wing with half the wing drooping helplessly to the ground. We did not have one airplane that could fly on the 13th.

But how the fortunes of war can turn. Miraculously, our engineers got our strips in operable condition by noon. That afternoon the Ferry Command started flying in brand new B-25-J replacements. This continued, and by the 15th our aircraft were back at full strength and we resumed flying missions.

This was a night and experience I will never ever forget. I now have a better understanding of what the British went through when the Nazis were hammering them in 1941. And what the Germans experienced in 1944–1945 under our round-the-clock bombing.

Finally, we can all be eternally grateful that the offensive death and destruction during the "great war" never reached our American shores.[1]

About this day, George Wells says:

"This was my longest day. It began about 4:30 a.m. when the Germans bombed us on Corsica. When the sirens blew we jumped out of our bedrolls and into our slit trenches. We then knew we hadn't dug them nearly deep enough. The raid lasted about fifteen minutes. Our fuel storage, bomb storage and some airplanes blew up. Those minutes were like a day! The devastating explosions were extremely loud. When daylight came, it was really a terrible sight and it took a long time to get over the widespread destruction."[2]

About this day, Bill Chapman, in a May 18, 1944 V-Mail to his wife, wrote:

"Darling—This last week has been one of the toughest and yet one of the most interesting weeks I have ever spent. I've really worked like a dog and have got results far beyond anything I expected was possible. I wish I could write the details but I'll have to save it until after the war, I guess."

In *Catch-22* Heller, while not alluding to this bombing raid, quite obviously wove it into his own version of the disaster. In one of the book's most memorable episodes, Heller highlights Milo Minderbinder, a pilot in Yossarian's squadron. Milo—"my name is Milo Minderbinder and I am 27 years old"—had formed an international business syndicate that included as its members both allied nations as well as the German government. At one point, in order to save the syndicate from bankruptcy, he signs a contract with the enemy to strafe his own outfit.

Curiously, and with connections to Milo, the first aircraft, in reality, to appear that horrific May 13, appeared to be a "friendly." It was a Bristol Beaufighter, the British twin-engine night fighter. It was later surmised that the Germans had captured it, left its markings on, and used it to catch their targets off-guard. Its job was to fly in before the main attack force and drop lighted flares to illuminate the target, this time the sleeping 340th.

TALE #10
DAVID MERSHON SPEAKS: 487TH
THE SAGA OF 7-F *WILLIE*

During the latter part of World War II the 487th Squadron of the 340th Bomb Group became known as the "Dog Face" squadron. It is understood that Bill Mauldin, the well-known wartime soldier cartoonist, gave his approval for the use of his GI characters on the 487th's airplanes. (Bill Mauldin went on to win two Pulitzer Prizes for his cartoons, one in 1945 and one in 1959. For the one in 1945, Dog Faced Willie played the leading part).

Col. Willis F. Chapman, left.
Cartoonist for G.I. Joe: Major John E. Rapp, center.
Col. Robert D. Knapp, right.

At this time 7-F had a lot of missions on it. Crew Chief T/Sgt. Michael P. Tarkany had done an outstanding job of keeping it from becoming a war weary, and to keep it being used as a lead ship. It was the natural choice for 7-F to be named after Mauldin's leading character. This 7-F became *Dogface Willie*.

At the same time 7-F was named *Willie*, and nearing 100 missions, a major milestone was becoming a possibility for three combat crew members who came up through the other three squadrons of the 340th Group—a record for the highest number of bombing missions flown by a crew member in World War II. This involved T/Sgt. Robert L. Helferich, a top gunner for the 486th who was on his second tour and closing in on his 100th mission. There was Major George L. Wells, group operations and training officer, formerly a flight commander from the 488th. Then there was Major Fred E. Dyer, assistant group operations officer, formerly a flight commander in the 489th squadron. Both Wells and Dyer were in friendly competition to try to outdo each other while still on an extended first tour of combat without returning to the States for rest. They both were closing in on 100 missions.

When you consider the above facts, you can see that the saga of 7-F *Willie* would result in its being the airplane to bring the four squadrons and group headquarters of the 340th all together for one big public relations event: three crew members each flying their 100th combat mission. Wells and Dyer couldn't resist the opportunity. They had never flown a combat mission together in the same ship, but had been on the same mission in different ships a number of times, including a special six-ship flight they flew on for some low-level missions.

Colonel Willis Chapman, the group commander, very reluctantly agreed to let Wells and Dyer fly together on their 100th, but only if the target was not considered heavily defended. That opened the door for Helferich from the 486th to fly as top turret gunner for his 100th. With Willie also in the act, the mission would have to be flown one day when the 487th would be scheduled to lead the Group.

Wells and Dyer both wanted to fly as pilot on the mission, so they tossed for it. Wells won the toss; Dyer would fly in the right seat and act as formation commander, while the group bombing officer Capt. Vincent (Chief) Myers, formerly lead bombardier in the 488th, would be the bombardier for the mission. "Chief" Myers was one of World War II's greatest bombardiers and is a legend in his own right. Major Richard H. Nash pushed to fly as the tail gunner on this mission. This now meant three majors and one captain all from Group staff on the same mission in the same airplane—more concern for Col. Chapman, but when Chapman decided to do something he backed it all the way. The 487th's top navigator, 1st Lt. Vernon (John) Lyle, was selected to fly as lead navigator. The crew was rounded out with another 487th man, highly experienced radio gunner T/ Sgt. John S. Wisanowski. The combined experience for the crewmembers in that one aircraft was over 530 missions (not including the 100 missions of the aircraft) covering the time period from when the 340th went into combat in North Africa to the day the mission was flown on Feb. 28, 1945.

Col. Chapman would have had a difficult time with higher headquarters if the airplane had had a mishap, either from combat or otherwise, what with three crew members all going for the top number of missions in World War II, and four of the crew members being from Group Headquarters with only 10 combat crew members assigned to headquarters.

Although the mission was deep in the Brenner Pass (3 hours and 25 minutes of flying time) it had flak of no consequence and no enemy fighters. The mission was successful. *Willie* came through and purred like a cat, just like a great B-25 was supposed to.

The day before the mission was actually flown, Wells and Dyer were told by a newspaper reporter that a pilot in the 15th Air force had flown 101 combat missions in two tours, one in the South Pacific. So they agreed, with Col. Chapman's approval, that they would both fly two more for a joint record of 102. Dyer flew his 101st mission as formation leader on a target in distant Austria. The plane received severe flak damage and was forced to land on the east coast of Italy. Several days later, Wells flew his 101st as formation leader, again on a target deep in the Brenner Pass. The flak was intense and the plane was hit over the target. They lost the right engine, forcing them to fly down the Pass to get out of the high mountains, picking up more flak and with more than two hours on single-engine flying before recovering at home base.

Just think what Chapman would have gone through waiting for the return of that 100th mission of *Willie* if either of those two events had occurred on that mission!

(George savors, "My best moment? The sighting of the beautiful island of Corsica ahead of us after a long and arduous mission and especially upon cutting the engines after completing successfully my 100th mission with a lot of news media and high-ranking officers all assembled there for the event.")

Wells and Dyer went on to fly a relatively uneventful 102nd mission. Col. Chapman than arranged for the three—Helferich, Wells, and Dyer—to return to the States in mid-April for a rest leave and then return to the 340th

Group. But the war in Europe ended while they were at home.

Dogface Willie (7-F) continued to fly and survive. It finished the war as a great lead aircraft in the 487th Bomb Squadron.[1]

MISSION #100: FEB. 28, 1945

With 488th plane as lead pilot of a 42-ship formation. Target was Salorno embankment on the Brenner rail line. Major Dyer was my copilot flying his 100th mission. He was acting as formation commander.

(They flipped a coin to see who would be the pilot, since they both wanted to be pilot . . . George won.)

"I had a gunner, Sgt. Helferich, who was also flying his 100th mission. Capt. Myers (Chief) was my bombardier. Maj. Nash was my tail gunner. Had lots of pictures taken when we landed."

Major George L. Wells, center, and Major Fred W. Dyer, right, just after completing their 100th mission, February, 1945. Offering congratulations is Colonel Willis F. Chapman.

Flying their 100th mission: George L. Wells, Fred E. Dyer, and Robert L. Helferich. Flying its 100th mission also: Dogface Willie (7-F), the dependable B-25 aircraft.

Col. Willis F. Chapman	Group Commander (in middle)
Maj. George L Wells	Pilot (left top of crew)
Lt. Vernon J. Lyle	Navigator (second from left)
Capt. Vincent "Chief" Myer	Bombardier (third from left, not seen in photo)
T/Sgt Robert Helferich	Top Turret Gunner (fourth from left)
Maj. Dick Nash	Tail Gunner (left in front)
Maj. Fred E. Dyer	Formation Commander (on right in front)
T/Sgt John T. Wisanowski	Radio Operator-Gunner (right of the ladder)
T/Sgt Michael Tarkant	Crew Chief

George with mission book

MISSION #101: MARCH 13, 1945

Formation commander with 488th on Aldeno railroad. Fill in the Brenner Line. Had lots of flak and had an oil line hit in the right engine and had to go on single engine over the target. We had to drop out of formation but we managed to get back OK.

This was the 1st medium bomber that has ever returned from the Brenner Line on single engine. We were on single engine for 2 hours. We were able to hold it at around 6500 feet. We were shot at again crossing the Po Valley by 40 and 20 mm. We then had a hard time getting over the mountains between Po Valley and the coast.

MISSION #102: MARCH 19, 1945

Command pilot with 489th on March 19, 1945. This was the Group's 800th mission. The target was a railroad bridge at Muhldorf, Austria.

This should be the last mission I'll fly before going home for a thirty-day leave.

The B-25 and her companions needed each other. It is not a stretch to understand how tight became the bonds between an animate and an inanimate object.

The B-25 was the most state-of-the-art mid-range bomber of its time. This powerful aircraft carried its crew from homeland soil toward the intensity of a war raging on foreign turf, an area of the world on which most of these men had never even dreamed of placing a foot. As they climbed on board, each one gambled his life in this aircraft, day after day, mission after mission. They depended on it entirely for their passage from home base to their perilous goal, often through terrible onslaughts of firepower and flak, and then to return, in whatever condition, whole or broken, back "home." To the best of its ability, this aircraft protected these men with a steely dependability, without hesitation or complaint, in doing whatever they asked of it. Sometimes it returned its precious cargo with its skin shredded, with a useless engine and a feathered propeller, with a tail half gone or its gas totally gone. The men climbed or were carried out.

Then the other half of the B-25's life took over. Her crew maintained her, healed her wounds, and nursed her every need. They sent her out whole, anxiously awaited her return, and ministered to her injuries. She belonged to them.

The following two stories give a glimpse of these attachments, these emotions—this, in truth, love.

TALE #11

FRANK B. DEAN SPEAKS: 380TH GROUND CREW
A LADY LEAVES THE 310TH

This month I lost a sweetheart. Along with some battle weary B-25Cs, Tissy Prissle was chosen for transfer. She was one of the great loves of my Army career.

By some strange metamorphosis an inanimate machine assumed a feminine personality, and we of the ground crew became her slaves. We showered her with our attention and affection. We existed to serve her. Her every need and want required our immediate pampering of her whims and appetites. We tenderly ministered to her ills and injuries, and grieved over her hurts. We became jealous and possessive, and, though we were hers, she was not ours alone.

We shared her with the combat crews who became fickle lovers. It was they that carried her away for a brief affair. They picked her up at her finest and returned her draggled, battered, filthy, and many times severely injured. They abandoned her on our doorstep. And left us to sit up with her through the night to provide the loving care necessary for her well-being. It was we who saw that she took her gas and oil, and saw that bandages were applied to her tender skin.

The combat crews came and went. Some transferred to other crews, some were wounded, some were killed. Others bailed out of aircraft that staggered back to the base. First pilots departed and copilots became first pilots on other aircraft. Gunners and bombardiers moved or rotated. Their allegiance to her lasted only as long as their assignment. But however much they loved her, we loved her more.

On our final day together Drake, Honig, and I groomed old *Tissy Prissle* for the last time. There was that final polish to the windshield and windows, our final time to preflight and run her engines, the last time we would top off her gas tanks. We waited for her to depart.

She tripped down the runway like a graceful young lady, and then daintily tucked her landing legs out of sight. With the grace of a ballet dancer she lifted her right wing and tiptoed around the field to allow her lovers a final look of admiration.

But this was only one side of her complex personality. She had fought like a dance hall hoyden, scratching, tearing, rending, and sending to smoking destruction those dainty German and Italian fighters that menaced those who were a part of her life. She was permissive as well as promiscuous.

For those who caressed her control wheels, fondled her bomb toggle switches, and squeezed the spade grips of her guns, she acquiesced to their every desire. Without hesitation she allowed herself to be pushed through black clouds of flak bursts and flaming necklaces of small arms fire. She served only to please.

She had fought and protected her scared and wounded lovers with a mother's devotion. With pulsating engines she had shoved through and around obstacles in an effort to get her injured home to safety. She had snarled and slashed at enemy fighters that tried to impede her progress. She had pressed through skies full of anti-aircraft fire to haul her wounded home.

It is small wonder that we called her "her" and she loved it.

Tissy Prissle had lost her youth in the skies of Africa, Sicily, and Italy, and was now like an aging actress retiring from the spotlight on the stage while a young, new replacement waited in the wings. She circled the field once again, proudly displaying row after row of miniature bombs awarded like medals for perilous trips. She looked just as graceful as ever, but distance created the illusion. We could not now see the peeling paint, or the metal patches that hid her flak and bullet holes. She headed south, and, with a saucy flounce of her empennage, she flew out of our lives.

I stood with misty eyes and watched her grow smaller and smaller until there was only the empty blue of the sky. Whatever anyone else thought, I knew her as a very gallant lady.[1]

CHAPTER 18

TALE #12
OLEN BERRY SPEAKS: 489TH GROUND CREW
BRIEFING TIME (9-D)

She was my plane. Yes, Joe Moore of Lakeland, Ohio (who named her) was crew chief, but I was his mechanic and I was with 9-D every day, almost, since she was commissioned until we returned to the States from Rimini, Italy. I feel that I knew 9-D better than anyone except Joe Moore.

Almost from the beginning, 9-D flew in the lead position. Of course credit for this must go to her flying crews, but Joe and I had to keep her in top condition.

I recall that some of the men who piloted this plane were Bus Taylor, Fred Dyer, Len Kaufmann, and Bill Chapman. She was a steady, well-balanced platform-type that bombardiers loved. The Norden bombsight and Shoran equipment were at home in 9-D.

9-D never came home with a feathered prop, never aborted a mission, and never had an injured crewmember. Truly, she was a great plane. As I recall, 9-D logged over 1,200 flying hours and had only three engine changes. (She flew 126 missions without a return due to maintenance.)

As you can tell, I loved that B-25.[1]

The following photos were taken of *Briefing Time* in 2008 where she resides. This is a B-25J Mitchell, renovated, restored, and refreshed at the Mid-Atlantic Air Museum in Reading, Pennsylvania as *Briefing Time*.

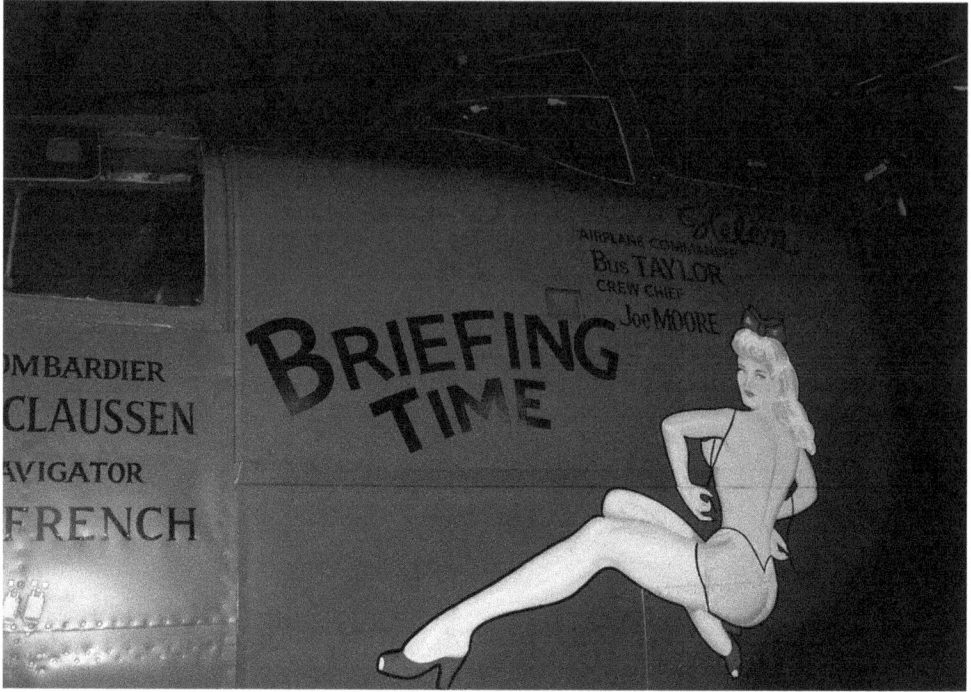

Briefing Time with its nose art beneath the pilot's side of the cockpit.

Note the painted bombs for perilous trips plus one ship for the sinking of the Taranto.

Pilot's Panel

Bombardier/Navigator "Greenhouse"

Guns of the Tail Gunner

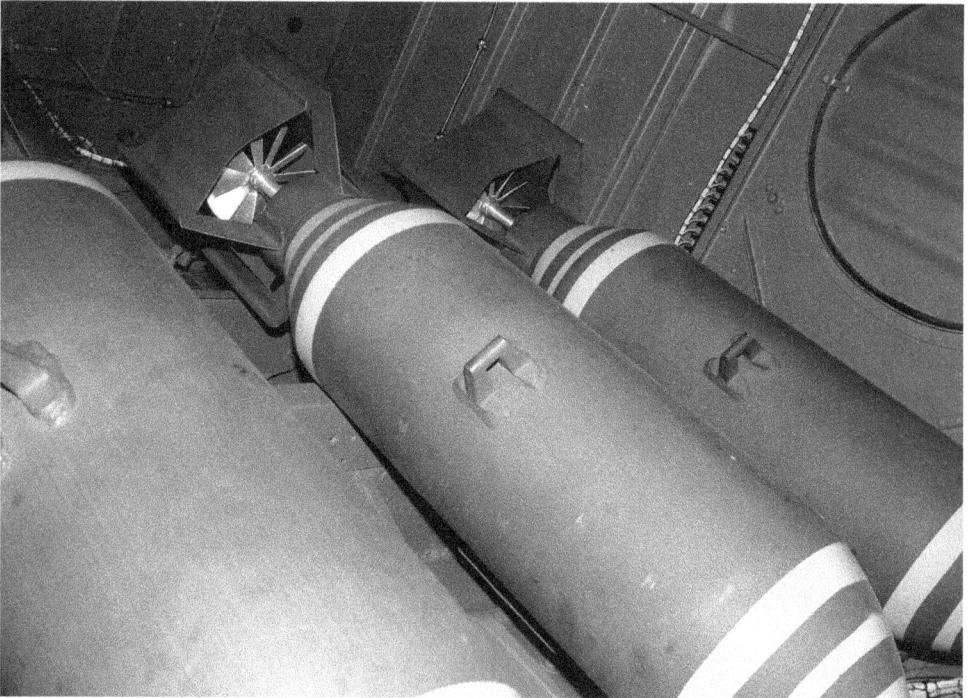

Bombs

INTERESTING FACTS
SOME COLORFUL MEMBERS
OF THE 57TH BOMB WING

DONALD K. (DEKE) SLAYTON:
WORLD WAR II PILOT

Deke was with the 340th (486th) in Corsica for 56 missions and with the 319th in the Pacific until the end of the war.

AFTER THE WAR: DEKE SLAYTON, NASA ASTRONAUT

Deke was one of the original seven United States astronauts. He was named as one of the Mercury astronauts in April 1959. He made his first space flight as Apollo docking module pilot of the Apollo-Soyuz Test Project (ASTP) mission, July 15–24, 1975—a joint space flight culminating in the first historical meeting in space between American astronauts and Soviet cosmonauts.

DITA BEARD:
WORLD WAR II RED CROSS GIRL

Dita was well known during her 13 months on Corsica. She was attractive and had a reputation for being able to swear and drink with the best of the men. She claimed that she "used to fly P-47s sitting on the pilot's lap."

AFTER THE WAR: **DITA DAVIS BEARD, ITT LOBBYIST**

Years later, during the Nixon Administration she gained substantial notoriety as an International Telephone and Telegraph (ITT) lobbyist. She wrote a memo, dated June 25, 1971, that went a long way towards proving that in return for hefty campaign contributions ($400,000) to the GOP, the Justice Department dropped its antitrust suit against the corporation. Syndicated columnist Jack Anderson received this memo, in which Beard also wrote that President Nixon and Attorney General John Mitchell were two of the few who knew of this contribution.

W. Mark Felt, number 3 official in the FBI and later of Deep Throat fame, was given this memo to authenticate. In spite of the White House's attempts to cover up the issues, the FBI released its finding of authenticity. Felt perceived that the direct request from President Nixon to back off the finding of authenticity was nothing less than pressure from the White House to cover up the ITT–GOP connection, pressure which Felt later characterized as "in some ways a prelude to Watergate." [Gentry, 2001, *pp. 716–717*; Woodward, 2005, *pp. 37–39*]

Dita Beard certainly stirred up the proverbial can of worms.[1]

LIEUTENANT JACK JOSEPH VALENTI: WORLD WAR II PILOT WITH THE 321ST

As a young pilot in the Army Air Corps in World War II, Lt. Valenti flew 51 combat missions as the pilot-commander of a B-25 attack bomber with the 12th Air Force in Italy. Honors earned included a Distinguished Flying Cross and five Air Medals.

AFTER THE WAR: **JACK VALENTI, LONGTIME PRESIDENT OF THE MOTION PICTURE ASSOCIATION OF AMERICA**

Jack led several lives after the war years. They included White House Special Assistant, movie industry leader, and author. He has a star on the Hollywood Walk of Fame.

NICHOLAS KATZENBACH: WORLD WAR II NAVIGATOR WITH 486TH BOMB SQUADRON OF THE 340TH IN NORTH AFRICA AND ITALY

He was shot down over the Mediterranean in 1943 and then spent 27 months in POW camps, escaping twice. He was recaptured both times.

AFTER THE WAR: NICHOLAS KATZENBACH, ATTORNEY GENERAL UNDER PRESIDENT LYNDON B. JOHNSON

In 1966 he became Under-Secretary of State and later Vice-President of I.B.M. Corporation.

LIEUTENANT JOSEPH HELLER:
WORLD WAR II WING BOMBARDIER

Heller was a bombardier with the 488th squadron who flew 60 combat missions in the twin-tailed B-25.

AFTER THE WAR: **JOSEPH HELLER, SATIRICAL NOVELIST, SHORT STORY WRITER, AND PLAYWRIGHT.**

He made literary history with the 1961 publication of his first novel, *Catch-22*, a book that has become an American classic. Told by an interviewer that he had never produced anything else as good as *Catch-22*, Heller famously responded, "Who has?"

DALE ARDEN: WORLD WAR II 488TH SQUADRON INSIGNIA

AFTER THE WAR: THIS DALE ARDEN SQUADRON INSIGNIA RESIDES IN GEORGE'S CLOSET.

George with Dale Arden: 488th Squadron Insignia.

He says, "The original 488th Squadron insignia was drawn by Alex Raymond who drew and wrote the comic 'Flash Gordon.' The girl is Dale Arden, Flash Gordon's girlfriend in the comic strip. This is an actual aircraft decal that was to be placed on an aircraft. It was given to me as a gift by Gordon Ainsworth, crew chief, after the war."

George with model of a B-25, a hand-built gift from a crewmate.

FINALLY...

In late April 1945 the 340th completed its final combat mission over enemy territory. Flying out of Rimini, Italy on the beautiful Adriatic Coast, the 340th saw the once powerful German Army in full retreat in Northern Italy following over two years of combat duty that commenced in Northern Africa. The war in Europe was soon to end.

As the years are drifting by the ranks of these war mates of the 57th Bomb Wing are thinning. The solid core of young, healthy, strong and dedicated men and women of the World War II era is now dealing with the tricks and traumas of aging. Joints ache and memories dim and the body will never again function as it did.

But take a look at those left as they, in their eighties and nineties, still prioritize their annual reunions. The eyes spark, smiles are quick, and mutual war remembrances abound. They are experiencing the entire swing of the pendulum of life—birth to death, war to peace, health to illness. Their lives helped to direct America's history. They have anchored its democratic freedoms. They were there for their country. And now, as they have always been, they are there, still, for each other.

Quick stepping, pop-eyed, and tuxedo-clothed Cody, his son's Boston Terrier, is George's devoted companion. Cody, with his short and widely spaced legs, possessively trots after, hops around, and sticks close as George heads for his La-Z-Boy.

This favored chair molds around George's familiar form, somewhat reminiscent of the intimate connection he had shared with his B-25 pilot's seat. In contrast, this La-Z-Boy is his long-time friend whose only function is to provide for his comfort and well-being. As he settles himself in it is apparent that there are few subjects George would rather discuss than those WWII years. Their intensity would imprint upon him forever.

And now, so many years later, he eases in to his aging years with his beloved wife, Shirley, and the quick-witted Cody. His attentive eyes move about his small home office area. Tucked in an alcove and topped with models of aircraft sits his wooden desk and its chair. On the wall behind that chair is a mix of simply framed news articles and photos, most in black and white: grinning George leaning out of the cockpit of his first trainer; his aircrafts; his buddies; and most heart-wrenchingly, the stop-frame recorded death of his friend "Red" Reichard. An aerial photographer had caught Red's exploding plane plunging its terrified crew to their fiery deaths while spewing out a long, slightly arcing, plume of grey-white, deadly smoke. George, seconds removed and with the ability to do absolutely nothing, was an eyewitness from his own endangered cockpit. This man, his story, and this frozen image are always with George.

Another wall holds his shadowbox-framed leather bomber jacket and his medals. Brought out from a bedroom closet, and carefully unwrapped, is a large original 1945 painting of squadron mascot Dale Arden, girlfriend of comic strip-hero Flash Gordon, which was to have been the nose art for a B-25. While George is a modest soul, the rip-roaring, bomb riding, buck-naked Dale, holding back a lightning bolt high above her raven-haired head, is not. Hence, her home-in-the-closet.

Like Ishi, George is the last of his tribe. He talks of the past, enjoys the present, and touches on the future. Were it possible, his wish would be that his and Shirley's final resting place be near that of Bill Chapman and his wife, Charlotte, at Arlington National Cemetery in Virginia, America's most treasured site for its defenders. However, choice of location on that hallowed ground is not an option. So when age, rather than enemy fire, dictates the time, his home's Fort Indian Town Gap National Cemetery, in Annville, Pennsylvania, will open its military arms and welcome its son-hero.

"As I said at the beginning of this book, the events and effects of WWII are still vivid in my mind. I remember the planes and the missions. I remember sleeping on the ground, the same poor food every day, using my helmet to take a bath in the cold weather, the continual loss of friends, flying long missions over Europe, the big Mt. Vesuvius eruption, and the German attack on Corsica that destroyed most of our airplanes and killed 22 (219 casualties).

"And when the war was over I remember the relief and joy that glowed on everyone's face. In fact, people were jubilant. There were people who cried for joy and those that cried for their lost relatives and friends (me too). I lost my original bombardier and a cousin and a lot of friends made during my long combat tour. It was relief and sadness all mixed in together".

But mostly, George remembers his friends, his crewmates, and the great dedication and friendships they enjoyed. He is surprised at being among the last of these men standing. He is being allowed to bid a respectful and loving farewell to each of these brave, honorable, and dedicated men, these friends and brothers with whom he is forever bonded.

And that, of course, includes you, Joe.

238

George saluting at Arlington
National Cemetery .

Box of Six, missing one in honor of
the deceased veteran.

TAPS

Day is done,
Gone the sun,
From the lakes,
From the hills,
From the sky.
All is well,
Safely rest,
God is nigh.

AFTERWORD

AUTHOR

The October sun is warm. The blue, cloudless sky, actually, is perfect. The black tarmac under my sneakered feet supports interesting activity as a variety of unusual aircraft arrive and take off in preparation for tomorrow's 2010 Air Show at Virginia's Culpepper Regional Airport. Voices are soft against the erratic chugs, snorts, and explosions of powerful engines as they bunch their muscles for flight.

And there she is. My goal. My focus. At the far end of this stretched out tarmac, all by herself, stands my B-25-J bomber. There is Corsica. There is the 340th Bomb Group. I feel the thrill manifesting on my skin as I stop and draw in the breath of one seeing such a long time and seldom visited dearest of friends. Hello, again, plane.

This North American B-25 was one of the most famous twin-engine bombers used during WWII. It was the single most heavily armed aircraft, some versions carrying as many as eighteen .50 caliber machine guns. This Mitchell bomber was named after the outspoken Gen. Billy Mitchell, who proved once and for all that bombers could destroy targets and that wars would nevermore be decided only on land or sea. The world turned its eyes to this aircraft when, on April 18, 1942, sixteen land-based B-25 Mitchell bombers, under the command of Lt. Col. James Doolittle, were launched from the Navy aircraft carrier USS *Hornet* and attacked 5 Japanese cities in an unfathomable and daring raid that brought attacks to the Japanese homeland for the first time in 2600 years. The plans for this (later dubbed) "Jimmy Doolittle Raid" were so tightly guarded that the men who volunteered for this "dangerous secret mission" were not informed of a single detail until they found themselves aboard the carrier.

I had made arrangements with the owner of this particular aging but restored B-25-J for a flight. A flight! Understand, I feel I know this aircraft. I know countless stories, I have read extensively, I know my father's personal experiences and opinions; I have models, parts, pieces and a great variety of mementos from my father's belongings about the B-25 throughout my home. As an artist, I have drawn this plane, in detail, on any number of occasions for various reasons.

But today is different. In front of me is the final piece to my puzzle. Here is my ride. My experience. I move in, around and under this plane again and again. I settle into the pilot's seat, then the copilot's area. I stand behind those seats where the navigator with his array of papers and maps had stood. I peer down the lonely tunnel the bombardier had to travel to reach his so highly visible nose cone. This tunnel has to be traversed either on your stomach or on your back, pushing and pulling your body along as your shoulders slide along its walls and your face is inches from the ceiling. I make a few attempts to experience this, but claustrophobia keeps me from completing it. I could not have been a bombardier!

I peer out the upper turret gun position, and then step to the waist gunner's windows. Placing my hands on the gunner's two grips and trigger I dwell on the heavy, fierce roll of the gun belts before me, my mind full to bursting of images. Hunched over, I proceed toward the tail gunner's position. This, momentarily, is a questionable journey because that already snug interior gets progressively more confining. This one, fortunately, I am able to handle and can experience the isolated tail gunner's remoteness from his crewmates by both distance and his back-facing direction.

The hour arrives and I, with four other passengers, climb the small metal ladders under the plan's belly to board.

I have brought three "friends." The first is my father's B-2 bomber jacket. As I slip it on, my hand goes to the right pocket to feel for the small box of NoDoz still resting there. Second, in the left pocket, I had placed his silver metal pilot's wings, and third, on my thumb sits his 1935 West Point ring. During the war the ruby was lost from that ring so Dad, temporarily, had substituted. He carved the end of his plastic red toothbrush handle for a totally acceptable replacement.

Earlier, the limited space of the aircraft had taken me a bit by surprise. A 6-foot-tall crewmember would have had to stoop to navigate. Bodies passing each other would require twisting and curling, with heavy combat boots vying for precious floor space.

I slide into my seat, tighten the harness with its unusual and now long-obsolete hooking system, and adjust my headset. I inhale the scent of oil, canvas, and metal. Preparation is under way.

Shortly those famous Wright R-2600-13 Double Cyclone fourteen-cylinder air-cooled radial twin engines turn over, and, with thunder and roar, breathe life. The plane rumbles and vibrates, demanding attention.

This is why I am here.

I can "see" George Wells. His hands are adjusting his headphone over his light hair, then gripping either side of the figure-8-type yoke as he settles into the plane's left first pilot's seat, readying for the designated mission. This one is not to be a "milk run." To his right, like a hand in a glove, eases Fred Dyer as the Formation Commander. With calm deliberation he checks his instruments. He glances out the window, fingers absently touching his small dark moustache. Ruggedly handsome bombardier Vincent "Chief" Myers has removed his parachute, allowing his large shoulders to squeeze into the bombardier's area, and is now making adjustments to the precious Norden bombsight.

At takeoff the B-25 owns the runway with its powerfully increasing speed and screaming power, until it parts company with the earth into the now-welcoming and quiet skies. I see the turret gunner, the waist gunner, and the tail gunner as they are intently scrambling about in this cramped space and, once under way, occasionally firing short bursts to test their guns. There is a quiet tension and nervousness as these vital preparations are checked off. I can feel the anxious strain that permeates this B-25. But I can also feel the confidence of these men who, with amazing bravery, stepped up for us.

My ghost crew crosses over the pristine waters of the Mediterranean Sea to the Italian countryside so similar to the Virginia farmland now beneath us. I watch these long-ago warriors reach, then veer from, the IP and settle on that direct course to their target. Nerves strain to snapping as their path is locked into that three- to five-minute deadly and unswerving bomb run. "Bombs Away!" Then, in an instant, this aircraft is forced to perform to its utmost in evasive maneuvering, banking, climbing, dropping, and twisting to escape intense, accurate, and violent bursts of gunfire and splattered, razor-sharp chunks of jagged, ripping, slicing, metal flak. The fate of this B-25 and its crew is unknown for endless moments.

"I'm the bombardier," Yossarian cried back at him. "I'm the bombardier. I'm all right.

"I'm all right."

APPENDIX A:
GEORGE'S MISSION BOOK

George L. Wells flew a record-breaking 102 bombing missions.

George with mission log

MISSION #1: OCT. 26, 1943

First mission, Tues. Oct. 26, 1943 to bomb the town of Terracina, West Coast of Italy.

Oct. 27 Bad weather—mission cancelled
Oct. 28 Bad weather—mission cancelled
Oct. 29 Bad weather—mission cancelled
Oct. 30 Bad weather—mission cancelled
Oct. 31 24 B-25s & Kitty Hawk (P-40s) Fighters. Was #2 in 3rd box of airplanes. After flying to initial point, was called off due to weather. Target had been harbor at Ancona, East Coast of Italy

MISSION #2: NOV. 1, 1943

Bombed Ancona Harbor. Hit and sunk transport in harbor. Started fires on docks in the city. 4 hours & 15 minutes over & back. Over 150 miles in enemy territory. Flew up Adriatic Sea. Flew ship Oh-Daddy, which had a pretty girl painted on it. No. 2 in 2nd box. 36 B-25s and 12 P-40s. Had some ack-ack fired at us. This was the 1st target of this type for the 340th. Each plane carried 4,000 pounds of bombs. Later it was reported 488 sank the last German held Italian cruiser and other merchant ships, besides starting large fires in the dock area.

MISSION #3: NOV. 7, 1943

In Oh-Daddy, shipping in harbor of Ancona, high clouds over target. Finally found hole after flying over target for 18 minutes. Dropped bombs 4,000 lbs per ship— 36 ships (planes). #3 position in 2nd box and did flying even though copilot lost 3 planes this week. Lt Beebe killed. Lt. Jordon missing in action and Sgt. Milan seriously wounded. All 3 were in my shipments areas

MISSION #4: NOV. 12, 1943

To a Tatoi Airport, Athens, Greece—48 airplanes from 340th & 48 from 321st & 36 from 310th 82nd fighter Wing for support—P-38s, P-39s, P-40s & Spitfires. Never saw so many planes in the air at one time. Weather stopped us from going to original target. Started a run on alternate target on Airport at Berot, Albania. The Jerries (Germans) put up flak so thick you could have played baseball on it. All of the old flyers said it was the most ack-ack they had ever seen. Enemy Fighters were diving down on gun crews, the other squadrons dropped their bombs but we couldn't get near the target. As yet I don't know whether we got credit for the Mission or not. Had 12 frog (smaller) bombs per plane. Finally got credit for mission.

MISSION #5: NOV. 14, 1943

Sofia, Bulgaria. This was the first time Bulgaria was bombed by the Allies—48 B-25s from 321st, 48 from 340th, 48 P-38s for escort. It was a five-hour mission at 18,000 feet. Had loads of ack-ack and ME (Messerschmitt) 109s fighting our planes during the run over the target. I saw three 109s go down. 270,000 lbs. of bombs dropped on railroad yards and warehouses. Saw trains running out of yards as we came in on them. It was a thrilling ride! Seven Jerries shot down and one of our own.

MISSION #6: NOV. 15, 1943

Airfield at Athens, Greece. 48 340th, 48 P-38s Frog bombs dropped and total airfield covered by fire. Enemy planes took off but never caught us. P-38s kept them away. Mission very long at 13,000 feet & above. No oxygen, which made us very tired. Had to sweat out gas the whole trip. Had 50 mph headwind coming back. 6 planes didn't make it and had to land at different places. No one hurt. We landed with only 10 gallons in our right tank, not enough to make another circle of the field. All planes had fuel shortages and damage, so everyone wanted to be first! O'Leary, Sweeney, MacDonegally went down over the target at Athens, six men aboard; 4 chutes seen to open as plane went down.

NOV. 25—THANKSGIVING DAY

Moved to Foggia, Italy. Had wonderful dinner.

MISSION #7: NOV. 26, 1943

Enemy positions at Fossacesia, Italy. 12 1325s. We escorted close support mission (strafing enemy lines). Flew #2 position 1 hour 30 minutes. Turned off target at well over 300 mph—low to ground

MISSION #8: NOV. 27, 1943

Lt. Saylor let me fly from pilot's seat though I was scheduled as copilot. Bombed shipping at Sibenik, Yugoslavia. 24 B-25s unescorted.

MISSION #9: NOV. 29, 1943

Flew copilot for E.J. Smith. Supposed to bomb Tronto but weather kept us from seeing the target. Then we flew to east coast, we bombed Giulianova, Italy. Bombed bridges and marshalling yards. Ran into a lot of flak as we came off the target. Must have been 105mm because they didn't fire straight up. 33 B-25s unescorted.

MISSION #10: DEC. 1, 1943

With R.M. Johnston. Enemy positions near Casino, Italy. Took off in morning but called back on account of weather. Take off again in afternoon. Johnston couldn't stay in formation and went over target by ourselves! 5,000 lbs. of bombs. Did not get credit for mission because bombs were not dropped. Johnston had to go around on coming in for landing. Later got credit. Dec. 3rd.

MISSIONS #11, #12: DEC. 2, 1943

Same target. 2 missions this day on enemy position at Casino Abbey with Whitehead & E.J. Smith. Had something hit the pilot's windshield but did not come through.

MISSION #13:

To Sibenik, Yugoslavia. Full bomb load. 24 planes, target shipping, docks and marshalling yards.

MISSION #14: DEC. 5, 1943

Was very sick but flew anyway. Copilot for Dean on raid to Aguilla. Bad weather made us bring our bombs back but ran into a lot of ack-ack on coast. Couldn't get left engine out of high blower.

MISSION #15: DEC. 7, 1943 (1ST AIR MEDAL)

Copilot for Dean. Town of Pescara. 5,000 lbs. per plane, more ack-ack on this mission than any of the others. My plane hit in 2 places, bomb door and in radio compartment. (Air Medal)

MISSION #16: DEC. 16, 1943

Zara Port, Yugoslavia. Marshalling yards, docks, ships. Copilot for Dean, 27 B-25s, 5,000 lbs. per plane. Small ack-ack.

MISSION #17: DEC. 17, 1943

Terni, Italy. Had to turn back because of bad weather. Made 360 degree turn over enemy territory. Flew with Dean, tried to find alternate target. Enemy opened up on us but didn't do much damage. Had spitfires for escort.

DEC. 19, 1943

Made first pilot! Given plane named Skunk Hunter 8P (# of plane) with Brooks, La Pointe, Sgt. Wood as enlisted crew.

MISSION #18: DEC. 30, 1943

Flew #3 position as 1st pilot. Bombed marshaling yards at Falconara, Italy, 5,000 lbs. per ship. 27 from 340th, 27 from 321stt—ack-ack.

JAN. 3, 1944

Moved to Pompeii Field near Naples and Mt. Vesuvius.

MISSION #19: JAN. 13, 1944

Leader of 2nd element. Major Cassada was copilot. Airfield at Guidonia, Italy. Plane was hit in wing. All planes in flight were hit. This was first time to wear flak suits. Red had ack-ack go right through his map while he was reading it. Had more ack-ack on this raid than any previous.

MISSION #20: JAN. 14, 1944

Flew as leader of 2nd element on Pontercorve Bridge, Italy. My plane had 15 holes in it, seven of which were in the right engine nacelle (cover). The left engine was hit and leaking oil. The right main gear tire had been hit and blown out, leaving me with no tire or brakes for the right wheel. Other hits in wing and bombarding compartment. I made a good approach to landing but after I touched the ground, I had my hands full. When I finally got the plane stopped, we were facing the opposite direction! The group lost 2 ships on raid.

I've been sick with yellow jaundice for the past week. Just found out that my plane had 33 holes in the raid on the 14th. The hydraulic system was shot out in the right nacelle.

MISSION #21: JAN. 16, 1944

On Terni, Italy. Flew lead of 2nd element. Easy mission. Ack-ack didn't get anywhere near us.

MISSION #22: JAN. 18, 1944

On Aquaduct at Terni, Italy. Led 2nd element of lead flight. Lead ship dropped out in return and I took over the group and brought them back to field. Had spitfire escort. Ran into ack-ack on way back.

MISSION #23: JAN. 19, 1944

On Rieti Airfield, Italy. Flew as copilot with Dean as lead ship of the box. Have good chance of becoming flight commander, but by the time I get it, I'll have too many missions, which won't leave me enough time to serve out the 3-month period required to get a captaincy.

MISSION #24: JAN. 20, 1944

On Viaduct near Avezzano, Italy. Lead ship in 3rd box. Flight leader of 2nd Flight. Plane hit by ack-ack on return.

MISSION #25: JAN. 21, 1944

Mission on marshalling yards at Avezzano, Italy. Leader of 2nd element of 1st box of 1st flight. Four enemy fighters tried to come in but Spits shot one down and kept the others away. Ack-ack was very light.

MISSION #26: JAN. 22, 1944

Invasion force landed near Rome. I led 4th flight and blew a cylinder head on take off. Engine kept cutting out but cleared up enough for me to go on to target. Target was road and railroad junction at Colloferro, Italy near Rome. Had ack-ack going in and coming off target.

MISSION #27: JAN. 29, 1944

Flight leader for squadron on marshalling yards and docks at San Benedetto, Italy. Couldn't get to original target because of overcast.

MISSION #28: JAN. 30, 1944

Target: road junction at Frascati, Italy. Flew pilot in #5 position. One of our pilots made a good job of landing with one engine and only one wheel. Only one person was hurt in the landing although the plane was a total wreck.

MISSION #29: FEB. 2, 1944

Road junction at Marino, Italy. Lead 2nd element of 1st Squadron box. Some ack-ack.

MISSION #30: FEB. 5, 1944

Marshalling yards at Orte. Bad weather forced us to bring our bombs back. Flew in #5 position. Found out that we were just sent up to get enemy fighters in the air so our own fighters could jump them.

MISSION #31: FEB. 6, 1944

On marshalling yards at Frascati, south of Rome. The whole town was practically blown up on this raid. We flew nearly over Rome when coming off the target. Plane was hit but no damage done. Flew in #3 position. We lost a ship and crew today.

MISSION #32: FEB. 7, 1944

On Cisterno, north of Rome. Lead 2nd element. Fighters jumped us but Spits took care of them. Boy it was cold. 20 below zero. (Heaters were turned off in the planes due to fear of starting fires during combat).

MISSION #33: FEB. 8, 1944

Cisterno, Italy. Flew lead of 2nd element. We bombed the whole town because the Jerries had troops and tanks dug in around the buildings.

MISSION #34: FEB. 10, 1944

Mission on Lanuvia, Italy. Target to flatten the town on the invasion bomb line. Bad weather over target. Flew 2nd element leader in 2nd box.

MISSION #35: FEB. 13, 1944

Mission on enemy positions (trucks, munitions, etc.) on invasion bomb line south of Rome. Flew element leader of 2nd element of 1st box. Last 2 ships on raid. Bombardier, Lt. Shanken, was hit in leg and plane was hit a number of times. Right propeller had a big hole in it. My two wing planes were both hit badly and had to drop out of formation. Lead ship of my box went down. Everyone got home but one whose chute didn't open. A number of other men were hit by ack-ack. All but one ship was hit. (Distinguished Flying Cross Award.)

DISTINGUISHED FLYING CROSS

For extraordinary achievement while participating in aerial flight as pilot of a B-25 type aircraft. On 13 February 1944, Capt. Wells, in support of the Anzio Beachhead, Italy, led a formation against a heavy enemy troop, gun and supply concentration near Campoleone, Italy. While on the bomb run, the formation was met by a withering barrage of heavy, intense and accurate anti-aircraft fire. His plane was holed in many places including a burst in the left rudder which severed the cable and controls. Displaying superb flying ability, he managed to keep the formation intact, thereby enabling the bombardiers to drop their bombs in the target area with devastating effects. Unable to maintain the crippled aircraft in formation, he returned unescorted and landed safely at his own base. His aggressiveness, courage, and devotion to duty on this and many other combat missions have reflected great credit upon himself and the Armed Forces of the United States.

MISSION #36: FEB. 15, 1944 (4TH AIR MEDAL)

Over Abbey of Cassino. There wasn't a thing left standing after we went over the target. Flew lead of 2nd box. This abbey was a German stronghold holding back the 5th Army, US 1 hr. 40 min.

MISSION #37: FEB. 16, 1944

Mission: Campoleone, beachhead. Flight leader 2nd flight. Worst mission I've ever been on. Had direct hit right through tail and cut my rudder cables. Was undecided whether to bail out or not. We salvoed our hatch and was all ready to jump, but I finally decided to bring it in. We made it all right and I was never so thankful to hit the ground. The enlisted men wanted to jump but I talked them out of it. We lost one of our planes, "Red" Reichard and Dean went down with it. Also Lt. Dunaway who came over with us. Seven men in the ship and only 3 chutes were seen to open. The plane was hit in the right engine and blew it all apart. All the fellows think I'm lucky because today was the third time I've brought back a shot-up plane.

MISSION #38: FEB. 17, 1944

Mission on Supply Dump Southwest of Rome on Tiber River. Led 4th box—B-17s were getting all the ack-ack on this trip. Saw one crash into water. Flew over outskirts of Rome.

MISSION #39: FEB. 19, 1944

On enemy position on beachhead. Led Sqd. flight. Jumped by 12 fighters but gunners kept them out. Jerries threw up a wall of ack-ack, but was able to turn box inside of it and not one plane was hit. The 321st went in on target ahead of us and lost 8 ships to ack-ack and fighters.

MISSION #40: FEB. 20, 1944

Mission on Anzio beachhead. Led 340th and 321st Bomb Groups to target of enemy positions. The Jerries put up a wall of ack-ack and our flight got a couple of holes.

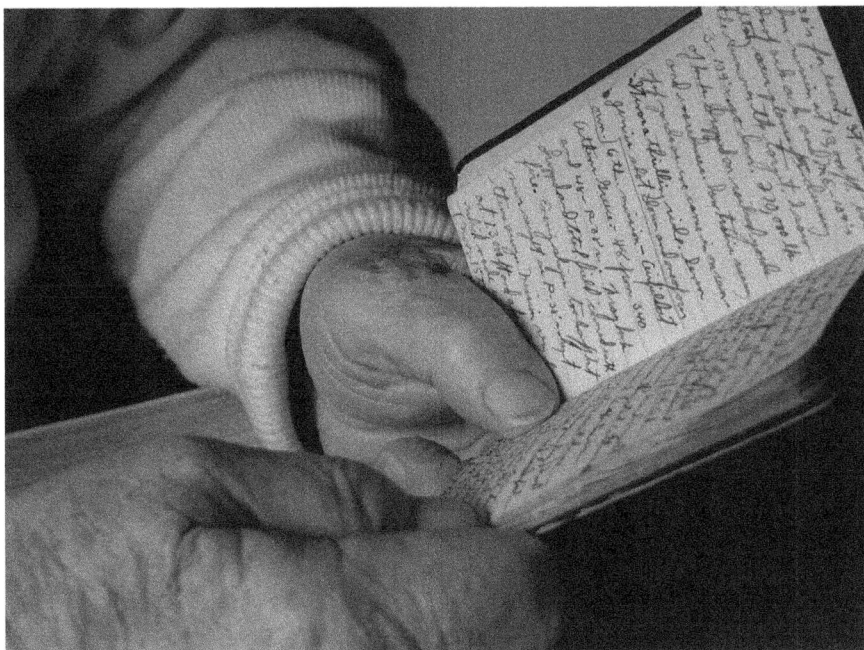

MISSION #41: FEB. 21, 1944

Mission on enemy positions on Anzio Beachhead. Led flight. Bombs not dropped. Target weathered in. 2 ships missing today.

MISSION #42: FEB. 22, 1944

Flight leader. Target marshalling yards at Foligno, Italy. 486th Sq. ran into flak and one man was killed. I turned our flight short of flak. We hit the center of yards and started big explosion and fires.

MISSION #43: FEB. 24, 1944

Mission on airfield at Fabrigo, Italy. Led Sqd. flight. Had a Col. Taylor for copilot from Tactical Air Force. We blew up a number of enemy fighters on the ground.

MISSION #44: FEB. 28, 1944

On landing strip at Comino, Italy. Led flight from 340th & 321st. Bad weather all the way. Clouds over target, unable to drop bombs. One man from 487th hit with flak.

MISSION #45: FEB. 29, 1944

Enemy positions on Anzio Beachhead. Led Sqd. flight. Lots of flak and a few enemy fighters.

MISSION #46: MARCH 2, 1944

Enemy positions on Anzio Beachhead. Led Sqd. flight. Lots of flak and a few enemy fighters. Saw a B-24 going down near target and 6 chutes open. Maj. Cassada flew as my copilot.

MISSION #47: MARCH 9, 1944

Shipping and docks at Porto San Stefano, Italy. Led Sqd. flight. Lot of ack-ack but 321st got all of it.

MISSION #48: MARCH 13, 1944

Flight leader. Target: marshalling yards at Perugia, Italy. Ack-ack was moderate. Fighters took off after us but we had a head start and they didn't catch us.

MISSION #49: MARCH 15, 1944

Led Sqd. flight of 12 planes on the town of Cassino. Our own troops were withdrawn from the city and the others are sending formations over there every 15 min. for the rest of the day. Our group was the first to bomb the town. The Group Co. flew as my copilot (Col. Bailey). The Group flew two missions over the target.

MISSION #50: MARCH 20, 1944

Led Sqd. flight on target at Perugia, but bad weather made us turn and bomb alt. at Terni. Vesuvius is really acting up and lava is running down the sides of the mountain.

MISSION #51: MARCH 28, 1944
Led flight to Perugia, Italy. Target: railroad bridge over river using 321st airplanes. Lots of ack-ack. Had seven holes in my ship and every ship in the formation was hit. We lost two ships from ack-ack. The 321st also went over the target after us and lost two ships. Promoted to Captain.1

MISSION #52: APRIL 1, 1944
Led group on my 52nd mission. Target: railroad bridge at Orvieto. Had the group CO Chapman as my copilot.

MISSION #53: APRIL 2, 1944
Led group, target railroad bridge NW of Orvieto. Had the Group operations officer as my copilot, Major Ruebel.

MISSION #54: APRIL 7, 1944 (6TH AIR MEDAL)
Led group on railroad bridge at Ficulle, Italy. Had Col. Chapman as my copilot, lots of ack-ack and Col. made us turn on the wrong target, so we had to do a 360 and come in again.1

MISSION #55: APRIL 18, 1944

Led group on mission. Target railroad bridge at Ficulle. Bad weather kept us from getting to the target area. Bombs weren't dropped.

On May 12, 1944, George flew his 56th mission. Here began the time frame of the events that eventually filtered into *Catch-22*. From this point on, until the completion of George's 102nd mission and war's ending, these official 340th Bomb Group War Diary entries have been matched - exact day to exact day - with George's mission entry. The intent is to compare the entry of a single pilot with the general more extensive entry of the Group's overall activities for that self-same day.

MISSION #56: MAY 12, 1944

We were told last night about the big push that started this morning in Italy. I led the group with 60 planes in the formation. The target was the town of Itri on the Adolph Hitler line. Gardner and crew went down over the target. At 9:30 p.m. the Jerries were over and bombed the field just north of us.

WAR DIARIES: MAY 12, 1944

The big push in Italy started today, but it was somewhat of an anti-climax for us. In the morning we had 60 planes out over Itri near Cassino, but cloud cover prevented bombing. Later in the day we trained on the same targets, a Jerry command post and a road junction. The results were not so good on the former, but good results were obtained on the roadblock. One ship of the 488th squadron went down like a rock in the heavy flak. Missing in action are pilots Gardner [Robert E. Gardner] and Powers [David R. Powers]; bombardier Rogers [Roy E. Rogers], and gunners Bradley [Samuel G. Bradley], Schmidt [Louis H. Schmidt, Jr.] and DeLucca [James C. DeLucca]. Almost all were very new men, and for the last named man it was his first combat ride...A few ground personnel have been ordered home for reasons of health, pressing personal needs or other reasons. The rotation system affecting ground personnel is open to criticism, it would seem. With one-half of one per cent of ground personnel supposed to be returned to the States monthly, healthy, efficient and generally reliable men are invariably passed up in favor of near-Section Eight cases [mentally ill], men in failing health, and men whose importance and value to the organization is nil, for one reason or another. One wonders why these men can't be sent home under other quotas.

MISSION #57: MAY 16, 1944

Led my Sqd. flight on Port of Piombino. Lots of flak but we dove around it and only got two holes.

WAR DIARIES: MAY 16, 1944

At 1000 hours today the men of Company "B", 845th Engineers completed our prefabricated S-2 – S-3 building. Identical to the one destroyed in the air raid, it is built veritably on the ashes of its prototype German AB-70 Cannister Bomb next to the old cork tree under which the first was built. In the way of office furniture the engineers also built us three sturdy tables. From headquarters we borrowed a typewriter for the joint use of S-2 and S-3. Target maps for our daily missions have been rushed down from Wing. Office supplies have been produced from nowhere. A few minutes after we moved from our temporary tent office to the new structure, S/Sgt. Sylvester Kepp of Elmhurst, Ill., and his assistant brought in another borrowed typewriter and proceeded to tie up all phones getting and consolidating and forwarding statistical information…The mission today to Piombino harbor was again very good…Major Robert Bachrach, former member of this organization who won the DFC with us in Tunisia for bringing home a smashed up B-25 with a dead pilot, was in today with the news he is going home on a 30-day leave. After serving with the 340th group as tactical inspector for several months he transferred to the Air Service Command, with headquarters in Naples…Good news continues to emanate from the Fifth Army front. American tanks advancing. Germans falling back to Pico and the Adolf Hitler line. The Gustav line seems to be pretty perforated…Headquarters men have a dance scheduled in Cervione tonight with local Corsican girls on the menu.

MISSION #58: MAY 21, 1944

Low level mission. 6 B-25s and 24 Spitfires. Bombed bridge at Sinalunga and machine-gunned railroad yards and German riflemen. Machine-gunned high tension lines and left them hanging on ground. Also started fires in the town.

DISTINGUISHED FLYING CROSS

For extraordinary achievement while participating in aerial flights as pilot of a B-25 type aircraft. On 21 May 1944, Capt. Wells voluntarily led a six-plane formation in a low level skip-bombing attack upon road bridges near Sinalunga, Italy. Displaying great courage and superior flying ability as he skillfully maneuvered at minimum altitude to the target area, Capt. Wells' precision-directed run enabled his bombers to skip their bombs into the embankments underneath several bridges, thereby blocking vital links in enemy communication lines. On more than fifty-eight combat missions his outstanding proficiency and steadfast devotion to duty have reflected great credit upon himself and the Armed Forces of the United States.

WAR DIARIES: MAY 21, 1944

Majors Bennett, Johnson, and Summers returned today after being "marooned" on Capri, where they went May 6 for a seven-day rest. The same morning Alesan was attacked by the Luftwaffe, German planes mined Naples harbor and the channel between Sorrento and Capri, making Naples Bay traffic impossible until the minesweepers could clean house. The officers managed to return to Naples May 18...Colonel Thomas of the Twelfth Air Service Command reported that the 340th has been given priority over all air force units in Corsica for food and supplies. Mused Major DeWitt F. Fields of Brilliant, Ala., group adjutant, "Are they doing it to bolster our morale after the beatings we took from Mt. Vesuvius and the Luftwaffe? Or, are they fattening us up for a move to another base, perhaps another theater?" Since the raid the number of 40 MM ack-ack guns has been doubled and 40 ack-ack machine guns have been put in....The twenty-first death in our group as a result of the raid was apparently the last fatality....Our group's T/O [Table of Organization] has been changed to allow 24 air crews per squadron and 20 airplanes, but the authorized number of ground crews remains at 16. What the hell? Possibly orderly room clerks will crew the other four planes...Ninety mile winds were reported over our operational area in Italy yesterday, canceling all operations but a suppertime low level attack by six planes on three bridges in the Sinalunga area caused some damage, by bombing and strafing.

MISSION #59: MAY 24, 1944

Led Group as 1st Pilot with 48 Ships in formation.   Target railroad bridge at Poggibonsi.
(See page 40, Heller's first mission.)

WAR DIARIES: MAY 24, 1944

Passes are now being given in headquarters and the squadrons again. S/ Sgt. Frank Tureck of Yonkers, N.Y., group supply clerk, returned last night from a three-day pass in Castelmare, Italy, and the charms of his Ida...Headquarters mess, largely because of the officers' predilection for being served by waiters, and the enlisted men's abhorrence of KP, is a good place to eat. These preferences and prejudices cost money, however; contributions for wages of Italian kitchen help being levied on all. Digging down for their francs everybody thought it would be a good idea to throw in a few more leaves and buy supplementary foodstuffs not handled by the Quartermaster. Hence once or twice in a week a B-25 hops to Catania, where in the city market Sgt. Ben "Heavy" Furstein of Brooklyn, chief cook, buys greens and fruits...Officers' club opened last night and successfully too despite the failure of the mess sergeant to provide ice cream as promised. The freezer was available, a home-made contraption put together by M/Sgt. Clay Hunn when the group was stationed at Catania, but the Quartermaster was unable to give the mess sergeant any ice. The situation at the Bastia ice house is temporarily critical...No mission went out yesterday, but three took off today to attack Poggibonsi rail bridge, Poggibonsi viaduct, and the Orvieto town road bridge. Nothing good was done with the first two assignments, chiefly because of cloud cover, but the 489th squadron gave a part of the Montepescali ammunition dump an effective high explosive treatment. Clouds hung over the Orvieto road bridge, so the 487th picked the railroad bridge north of it and claimed to have pranged it good, as the English say.

MISSION #60: MAY 27, 1944

Flew as copilot in lead ship as formation commander.   A captain or above now has to ride as copilot in the lead ship.   Target was railroad and highway bridge at Piettasanto.

MISSION #61: MAY 29TH, 1944
Target Viaduct and Bucine Bridge. Got permission to fly on the wing and flew #2 in second box

WAR DIARIES: MAY 29, 1944
Headquarters softball team lost a heartbreaker tonight to the 845th Aviation Engineers, the outfit who helped us back on our feet after the May 13 raid. Colonel McCrone's ball players made two runs in the 14th inning of a scheduled seven inning game and triumphed 5 to 3. Our disappointed worthies trooped into the enlisted men's club nearby, where the bartender of the evening, M/Sgt. Joe N. Kline, Coeur D'Alene, Ida., bombardier, plied them with gin, vino, and brandy, but at the regular rates...Equipment lost in the raid has been replaced rapidly...A C-3 (Speed Graphic) camera was issued to headquarters today to the joy of S/Sgt. Tom Smith, photographer laureate to the group. He takes many of the public relations and engineering pictures...Ballot application cards for voting in the primary and general elections in the States this year have been received in headquarters and many men are calling for them. The north and south viaducts at Bucine were hard hit by the 340th today.

MISSION #62: JUNE 1, 1944
Flew as formation commander with 487th. Sqd. target viaduct at Fossato. Not one bomb was dropped out of the target area.

WAR DIARIES: JUNE 1, 1944
Captain Anthony J. King, assistant S-2 officer, it is now revealed, was arrested in Naples about two weeks ago for not wearing his dog tags. The arresting M.P. was also scandalized at finding the culprit out of proper uniform and needing a shave. When the report of arrest finally filtered through channels yesterday, the group adjutant, Major Fields, forwarded the papers to Major Kisselman, S-2 officer, with the tongue-in-cheek remark, "Do whatever you think necessary." Put on the pan by Major Kisselman and the intelligence clerks, the Captain wailed: "It's time for us to pull out and go to France. They've got M.P.'s all over Naples and speed cops all over Corsica..." A report came in to S-2 that four Germans in American uniform are either known or thought to be hiding in the hills behind Alesan. They are equipped with binoculars and signal lights, according to the report. Matter is being referred to C.I.C.

[Counter Intelligence Corps] and French authorities...1st Lt. Clarence V. H[illegible], Quincy, Mass., assistant S-2 and the photo interpreter in the 488th squadron, is taking Captain Eggers' place as Group Photo Interpreter while the latter is in Cairo. Pressure from groups consisting of pilots and bombardiers from the squadrons, are trying to get him to pin down direct hits or otherwise report in detail the margin of accuracy or inaccuracy, but he cannily straddles the fence in cautious, marvelously neutral reports....Road bridges at Orte and Narni were objectives of the 340th today, as well as a viaduct at Fossato. The first two targets were well hit, and the Fossato installation was probably damaged too, according to the photo interpreter.

MISSION #63: JUNE 4, 1944

Flew as formation leader with the 489th. Led 36-ship formation. 26 Jerry fighters jumped us but Spits ran them off. 2 Spits were shot down out of the 12 we had with us. Target Gricigliano railroad bridge.

WAR DIARIES: JUNE 4, 1944

Capt. Eggers returned from Cairo last night where he had flown in a B-25 to spend a ten-day leave. Most happy to see him was S/Sgt. John Carraciolo, the S-2 clerk whose job it is to produce the bombfall plot for each mission. The sergeant had to work very late at night while Lt. Haynes of the 488th was substituting for Captain Eggers on photo interpretations, owing to the Lieutenant's unfamiliarity, or possibly lack of experience, with the work.....Colonel Charles D. Jones, our last C.O. before Colonel Chapman, who was shot down during an attack on the Littorio marshaling yards early in March, is a Prisoner of War it has been learned officially here at group headquarters. Everyone was delighted with the news, although for several weeks we were fairly sure he was alive in German hands. In one of her radio broadcasts the propagandist "Axis Sally" reported the capture of a 32-year old American colonel in Italy. We have learned that some of Sally's news can be trusted...Very good results were achieved by our crews today against rail bridges at Vernio and Gricigliano.

MISSION #64: JUNE 7, 1944

Target road bridge at Cecina. 1:45 hours. Flew as formation commander with 487th Sqd. German paratroops landed south of our field. Seven are still at large.

MISSION #65: JUNE 10, 1944

Flew with 489th Sqd. as formation commander on viaduct at Bucine, Italy. Saw one plane blow up and another crash.

WAR DIARIES: JUNE 10, 1944

Gradually more of our personnel wounded in the raid of May 13 are returning from the hospital. Only the more seriously injured for the most part are still undergoing treatment...One man in the 489th lost a leg, but is reported to be reasonably cheerful...Many of the combat "old timers" are going home or have already done so. Major Schreiner, 487th operations boss, has left on a permanent change of station, as has Major Parrish, 489th commander. Major Cassada, the 488th C.O., however, is expected back after 30 days rest at home...The group guardhouse, a modest cluster of two tents in the headquarters area, is now encircled by barbed wire, and gives the impression of taking on airs at this late date. The wire was idea of 1st Lt. Richard Kittay of New York, who seems to glow with pleasure when another prisoner [is given] to his charge. He proudly announced the other day that he had a record number of prisoners awaiting trial. This seems to be in keeping with the general spirit of ambition and expansion seizing the group. Squadrons are trying to get as many medals and decorations for their men as possible, and seem to be heroicizing many a routine action in the process. Each unit in the 340th group is trying to raise its bombing efficiency above the other, and the 340th is doing its best to outdistance the other two B-25 groups. Public relations, as a result of Colonel Chapman's interest, is trying to turn out more pictures and stories on personnel than it ever did before. Small wonder the Provost marshal has his eye peeled for more and more lawbreakers...There were four road bridge missions, which misfired because of a rack malfunction. Today our crews got excellent results on the Fano and Rimini marshaling yards, and also on Bucine south viaduct.

MISSION #66: JUNE 13, 1944

Flew as formation commander with 486th Sqd. Target: Perugia Road bridge. Lots of ack-ack but no damage.

WAR DIARIES: JUNE 13, 1944

A couple of days ago 1st Lt. Joseph Weil, bombardier on the plane co-piloted by Colonel Charles D. Jones, our C.O. before Colonel Chapman, returned here from the limbo of Missing in Action. Shot down with

Colonel Jones March 10th over Littorio marshaling yards (Rome), he is the only man who can give us any information as to what happened to the Colonel and the rest of the crew. Full details of Lt. Weil's experiences are not yet available, but it is known that Colonel Jones parachuted with the others and was taken prisoner of war (Headquarters was officially informed of the latter fact a few days ago). Lt. Weil landed near the target, handed his parachute to a startled Italian boy, shouted "Via Presto!" and made for cover. Soon he turned up in the city of Rome itself and received aid and shelter from sympathetic Italians, one of them a priest who gave him some money. According to Lt. Weil, Allied airman and many Allied ground troops, almost all evaders or escapees, are running around all over Rome staying out of reach of the enemy agents...A P-47 pilot who bailed out in the vicinity of Perugia about the time that the 487th Lts. Ashmore and Finney were shot down with their crew over Perugia, returned to Allied lines [illegible] with the story that Lt. Finney and others of his crew were safe and hiding out in German territory...Captain King, assistant S-2 officer, is being hospitalized for bursitis. "You've got housemaid knees, Tony!" his friends crow in delight.

MISSION #67: JUNE 22, 1944

Formation commander with 489th Sqd. Target: Viaduct at Gricigliana. Lots of ack-ack and fighters. 487th lost one plane and had another come in on single engine. Also a number of men were wounded. The 310th lost 3 ships over the target and had five crash landings.

WAR DIARIES: JUNE 22, 1944

Operations were a heller today for the 57th Bomb Wing. The 310th Group lost two ships and crews over Leghorn harbor, another over the Vernio north rail bridge near Florence, and two planes crash-landed at their base and ours. A pilot in another plane was killed and still another man lost an arm below the elbow. The flak was particularly deadly over Florence, which the 340th proceeded to fly over on the assumption it was an open city. We lost one ship to flak also, over Gricigliana rail bridge, a 487th plane piloted by 2nd Lt. Thomas V. Casey, and 2nd Lt Harry D. George. Other crewmen included 2nd Lt. Edward F. Dombrowski, bombardier; S/Sgt. Russel G. Ahlstrom, radio operator; Sgt. Paul M. Kaplan, turret gunner, and Sgt. George E. Obravatz, tail gunner....Three members of 2nd Lt. Richard L. Ellin's crew, which was shot down by flak May 25 near the Ficulle rail bridge, returned today with an absorbing story of their experiences. They are 2nd Lt. Garth B King, co-pilot; S/Sgt. Pete Vargo, bombardier, and T/Sgt. Henry E. Yocca, radio operator. They reported the pilot, Ellin, was killed in the crash, and the other two men in the crew taken prisoner. They themselves were seized by Italian fascists but when they were being transported in a car by two fascist guards a Spitfire strafed the road and the guards took to cover. As the plane buzzed away the trio slugged their captors and escaped to the refuge of friendly Italians. They were closely followed by fascists and practically a pitched battle resulted before three men were got to safety.

MISSION #68: JUNE 29, 1944

Flew formation commander with 488th Sqd. Target: railroad bridge at Cervo, Italy. I got my first look at French soil today because the target was very close to France. It looks as though we'll soon be working on Southern France.

MISSION #69: JULY 1, 1944

Flew as formation commander with the 489th Sqd. Target: tunnel north of Prato, Italy. Ran into flak to and from the target.

WAR DIARIES: JULY 1, 1944

The most vitamized of all liquids, thick foamy brew, equal in quality to anything in the States was sold to the boys today. Two twenty-ounce bottles per man at ten cents a bottle. This is the first Naples Beer received by this Group though it has been in production in Naples for several weeks now available to units making requests for it. Headquarters Enlisted Men's Club functioned as distributors. The club's normal take-in of $25.00 nightly correspondingly dropped when the boys had something other to purchase. Black-Jack was again the main sport of the evening with Big Joe Nichols, Georgia, continuing his unenvied uninterrupted losing streak and thus again being the principal contributor. He swore off gambling till the next evening....The Group was warned to prepare for a possible inspection by the General who was voyaging down to these parts to look things over. After feverish activity, each had their respective spots looking as meticulously neat as a tabernacle. The General was too busy to inspect...With no casualties, evidently proving the faultless quality of our new gas masks, Headquarters personnel were put through the customary-typed gas chamber. None of the lethal gases were used however....Cannetto

Railroad bridge and Tunnel mouth were today's targets. Obscured by rocky terrain permitting only a 15-second run resulted in bombs falling wide of target. The first 18 ships carried the new type narrow S.A.P. (Semi-Armor Piercing) bomb. Because of its slenderness, the plane can carry 6 x 1000 pound bombs rather than the usual load of 4 x 1000 pound load. General Knapp, Wing Commander, commenting on the concentrated pattern of this 6 x 1000 formation was highly pleased saying in effect that it is a concentration of such intensity that would leave without question the desired destruction of target when hit............

MISSION #70: JULY 11, 1944

Flew as formation commander with 487th Sqd. Target: railroad bridge at Ceva, Italy. I now get command pilot time when I fly as formation commander.

Notation from George: "Prince Bernhard of the Netherlands landed here today and I had to entertain him until the Col. (Chapman) arrived. He seemed like a darn nice fellow."

WAR DIARIES: JULY 11, 1944

Railroad bridges were again today's targets. One at Bistangno, Italy the other at O-276494. Each of the bridges received direct hits.... Prince Bernhard of Holland flew in as Co-Pilot today in a B-25. He has been in this theatre about three weeks and wished to visit a B-25 base.

Wearing a Royal Netherlands Officer's cap, self-assured confident bearing, he maintained throughout his visit a careful reserve. The colonel Willis Chapman, Jackson, Michigan, acted as host and guide. Lt. Blum [Actually, Wilbur T. Blume] of the 9th Combat Camera Unit, hurried over to operations, his C-3 Camera ready for action. Two attempts with flash shots failed, the Colonel chiding the now blushing Lt. Blum[e]. The Prince and Colonel Chapman then obligingly walked into the sun-light permitting another try. The objects by now become too camera conscious with final results unsatisfactory.

MISSION #71: JULY 12, 1944

Flew as formation commander with 489th Sqd. Bombed alternate target: railroad bridge at Chiavari, Italy. The primary target at Fenoch, Italy in the Po Valley was closed in by weather. Had one ship ditch. Five stayed with the plane and were picked up the next morning. Two bailed out and were never picked up.

WAR DIARIES: JULY 12, 1944

Foiled by smoke screen over the Ferrara Railroad bridge the first mission went to its alternate, the Zoagli Railroad bridge with satisfactory results.. The second mission met with excellent results over the Ferrara Road bridge. On this mission a photo ship lagging behind the formation was jumped by at least 12 Enemy fighters. The ship escaped unscathed but the gunners [Harold E. Winjum & Wallace E. MacRitche] have to their credit one destroyed and one probable. This same ship in the afternoon mission over Ferrara Railroad bridge, a repeat of the morning's mission, incurred motor trouble and was ditched out at sea on route home. [The author being somewhat removed from events got some details wrong. See Addendum at end of this Diary for more details on this incident.] The two gunners, the same gunners on the morning mission misunderstanding instructions bailed out rather than wait for ditching. Their chutes were seen to open and the men were seen to be afloat, but the men could not be later found by Air Sea Rescue who, through mistake, had been notified of this after considerable delay. The remaining five crew men however utilizing the dinghy were safely picked up..... At 2300 hrs Air Raid siren sounded off as warning of approaching enemy planes. Reported twenty miles out to sea. This was the first time we have been so alerted in over a month. T/Sgt. Ralph Wyland, S-2 section chief was again C.Q. as he was the night of May 12th when Jerry made literal hash of Ralph and the Airfield. Ralph delayed little in reaching the shelter constructed since the Raid..... Headquarters' softball team won again last night defeating the 845th Engineers, thus putting them in uncontested first place. By defeating the 489th Officers, the next scheduled game, they will cinch championship for the first half of the season. These games are usually very well attended. The diamond occupying the center of a rectangular space constituting the Headquarters' living area is in easy reach for all of HQ personnel to root for the home. The Officers usually bring out their beach chairs and file along the right field sideline watch-dogging to see that no injustice is done to the Headquarters gang.

MISSION #72: JULY 19, 1944

Flew with 487th as formation commander. Target: railroad bridge at Sassuola, Italy. Jumped by fighters but no damage.

WAR DIARIES: JULY 19, 1944

2nd Lt. William E. Blum [Should read: Wilbur T. Blume], C.O. of the 9th Combat Camera Detachment here is currently making a movie film about the Red Cross distribution of doughnuts and coffee to our crews after their missions. He has film footage of our formations going out to the target, the bombs dropping over the target, the target area covered by smoke and the men eating and drinking during the interrogation. Production of one scene showing the Red Cross girls actually handing out the victuals, was held up more than a week because the photogenic Red Cross girl was unavailable. Lt. Blum[e] obtained two good-looking Red Cross girls in Bastia by having them sent down here on detached service to film the sequence. Fraud! Fraud!..... In a few days he will start on a film depicting the various types of training undergone by 340th air crews in between missions [see page 288] ....Sgt. William E. Durkin of Buffalo, N.Y. Cyphers clerk is the second man from headquarters to leave on rotation. Overseas two years, Durkin was one of the noisiest and most aggressive "debaters" at the HQ Enlisted Men's bar, but also very well liked.... Apparently a big going away party is in store for him at the club.... A weather stations detachment of two officers and four enlisted men is now set up outside operations and services of the 340th with weather data before missions and also non-combat flights to other bases. The other night one of the sergeants came in S-2 and passed out cans of beer, he explained, by way of Navy men on a liberty ship in Ajaccio. How they got it does not bear investigation...... Leghorn and Ancona have been taken by Allied troops.....

MISSION #73: AUG. 1, 1944

Flew with 488th Sqd. as formation commander. Target: railroad bridge at Canneto, Italy. No escort and was jumped by 12 fighters. They didn't come in close because our 18 ships were in real close.

WAR DIARIES: AUG. 1, 1944

Payday... EM club again converted into Club Casino with both dice and black-jack the night's principal source of amusement..... Danger of the air, land or sea assault on this island has apparently subsided. The warning was not repeated and as a consequence the men have again become very lax and unconcerned. Vehicles along the road moving personnel from one location to the other no longer seem like a modern version of Jesse James' gang.... Cannetto RR bridge was again our target and again was successfully hit. Later recon reveal that bridge is no longer passable.

MISSION #74: AUG. 2, 1944

Flew with 489th Sqd. as command pilot. Target: railroad bridge just north of Nice, France. Plenty of ack-ack. This was the 1st mission for the group in France.

WAR DIARIES: AUG. 2, 1944

Captain Hass, MATAF, lectured to combat crews of the maize of techniques used by enemy interrogators to ween information from captured combat personnel. Confidentially said the Captain, every dirty trick used by the enemy is also utilized by our own PW interrogators.

General Orders, XXIIth Air Force announced that Colonel Willis F. Chapman, Group Commander, Captain Anthony J. King, Ass't S-2 Officer and veteran Navigator received the D.F.C. for outstanding achievement while in aerial flight over the enemy..... Captain King has been for the past few days back in the hills "Wild Boar Hunting"..... Today's target took us into the coastal fringe of southern France. What seems to be the first of the softening up activity preceding the long anticipated landing in that general area...... Flak over target ripped one wing of plane number 6E to bits, the plane went into a very steep dive. Lt. Hill, pilot, described the situation as four hands and four feet were pushing and pulling everything in the cockpit to bring the ship back to level. After a 4500 ft. drop the plane proceeded home, crippled but safely...... The target, Var River bridges #3 and #4 just west of Nice. No. 4 was well covered with bombs. No. 3 was missed.... Group Movie had a large turn-out to see Ginger Roger's masterful performance of "Lady in the Dark." Hollywood using the still untapped medium of Psychiatry to bring in the lush tempting qualities of lady Ginger, her dancing mastery and fantastically appealing scenery. The plot at no time was lost, confused or sacrificed by this extravagant artistry. The audience was quite pleased... HQ, softball team defeated the hard losing 845th Engineers by a 5-2 margin. They are now in a tie for first place. Play-offs will no doubt be arranged to determine area champs.

MISSION #75: AUG. 4, 1944

Flew as 1st pilot on a single ship weather reconnaissance over Italy and Southern France. Had no escort and we were lucky because we didn't run into a fighter. Shot up the docks at Nice. 1:50 hours.

WAR DIARIES: AUG. 4, 1944

S/Sgt. Warren Bickford, and Captain Kirk White both of Oklahoma were transferred back to the States on rotation. Both seemed very happy to bid fair adieu to their buddies..... Breakthru of Bradley's Forces resuscitated the many slumbering speculations regarding the War's end. Achievement of Britteny with its many port facilities would facilitate such influx of necessary man and matériel as to constitute an irresistible avalanche of power once it was properly grouped..... Again an attempt is made to harness the transportation facilities of this Group. Objection of the previous conditions was that the vehicles were not being properly cared for, that they were being used for personal reasons etc. The order brought into effect this morning requires each vehicle be turned into the motor pool by 0900 hrs each day, for a complete checkup at which time a dispatch will be issued to it for the day's usage.... Rail bridges in southern France were again the targets for today. The day's efforts again had excellent results with the two bridges intended as primary targets well covered and a Marshaling Yards, a target of opportunity, put out of commission.

MISSION #76: AUG. 6, 1944

Mission with 488th Sqd. as command pilot. Target: railroad bridge at Nice, France. Lots of ack-ack but made it home OK.

WAR DIARIES: AUG. 6, 1944

Sergeant Horace Samuel Deese, Pageland, S.C. Hard drinking, hardworking publications section clerk, met his brother Basil in Rome. Sgt. Deese who normally on the slightest pretense fills his flush face full of booze, celebrated this occasion having his brother, by no means a teetotaler, as companion. Their only recollection of Rome is through the end of a booze bottle..... M/Sgt. Paul McElroy, Phil., Pa, Group Sergeant Major, after one month of expecting the worse, received word that his youngest brother was killed in action on the second day of the Normandy landing. His second brother also in Normandy visited the youngest brother's organization. He was told that on the second night while on patrol he was ambushed with the entire patrol being wiped out............ A non-league game with 401 Ack-ack was indeed a sad display for the Headquarters gang. Their complete lack of interest or enthusiasm made the proceedings a mechanical bore. The 4-0 final score was no indication of how poorly the game was actually played... Two missions taking the boys for a 3 ó hour ride up the Rhone Valley and the second to the Var River bridges produced excellent results.....

MISSION #77: AUG. 9, 1944

Flew as formation commander with the 487th. Weather was extremely bad and kept us from going to the primary west of Borjac, France. We did get in on the alternate, which was the railroad bridge at Venti Migglia, Italy.

WAR DIARIES: AUG. 9, 1944

Thick overcast prevented bombing the Primary so the alternate target, Ventimiglia RR bridge, was bombed with excellent results...... Threatening rain all day finally materialized into a steady down-pour the entire evening and early night. This is the first rain in several weeks. Rains such as these brings on a nostalgia of the frequent summer night rains back home, then so detestable but now so sweet to remember..... Major Fields, Alabama accountant, Group Adjutant, speaking to Major Kisselman regarding a PR form he was asked to fill in that he did not want any publicity of the type that usually reaches home concerning a local boy overseas. After some discussion, the form was filled out and handed in..... That something big and unique is cooking for tomorrow is not only in the air but can also be felt in one's bones. Major Kisselman, Group Intelligence Officer and Major Ruebel, Group Operations Officer, were called up to Wing for a secret conference. Such secret conferences are not intended as stag parties. Upon their return three hours later they both looked as if the mouse had been swallowed...... Feature attraction for the day was a dog-fight between a piper-cub and a bi-winged 1927 modeled Wright plane some 1500 feet above operations. The interrogation of the then returning combat crews was momentarily suspended to watch this fight of the century. After living through several deaths the easily out maneuvered bi-plane having his foe again on his tail released in the way of a smoke screen the yellow anti-malaria powder it today was carrying. The cub won the applause of all spectators for its daring performance........

MISSION #78: AUG. 14, 1944

Flew with 489th as formation commander on coastal at Cap Camerat, France. Things really look as though the big invasion will come tomorrow.

WAR DIARIES: AUG. 14, 1944

The day before the invasion. Who said so? Well, when returning crews report seeing convoys of several hundred ships moving north, when Group S-2 receives bundles of top secret target charts and photos, when the Group is committed to send up at daybreak 72 ships and 54 more in the afternoon, called into secret session by the Colonel with several ranking visitors who seem to know a good bit more than their appearance would justify, when another secret session is called for all combat crews and when the briefing room with all the top secret drawings brought down from higher HQ is locked and kept under heavy guard should all be a least fair evidence that another tea party is in the air for the Jerries..... Captain King, Ass't S-2 Officer, Major Kisselman, S-2 Officer with two clerks, T/Sgt. Hickey Group PRO, and S/Sgt. Thomas H. Smith worked until 2 AM preparing briefing putting together necessary photos, maps etc. The office was however as calm as before any storm. Needless to say, each and every man is keyed for the situation. They have long expected it and now are ready for the eventualities that follow. To those members of this organization who were present to see the beginning offensive breaking the Mareth line, the fall of Cape Bon, the Invasion of Sicily, the Salerno beachhead, the Break-thru at Cassino etc were in no position to be overdone by this climax to sixteen months of operations. To all however, veterans and neophytes alike, especially in the light of General Eisenhower's words to the troops in Northern France that this is the critical week of this war, bore but one thought in mind, HOME, the sooner the better..... Preparatory to tomorrow's operations was the continuation of our endeavors of the past few days, the bombing of still more gun emplacements along the Marseilles' area. Three of the four designated areas, confirmed by photo coverage, were well hit with the fourth very probably hit...... The evening was spent with M/Sgt. Joseph N. Kline, Group Staff Bombardier nervously staying up the entire night and with the ground personnel gathering about in Thomas H. Smith little circles speculating. The Kitchen will open for business at 0230 hrs rather than the usual 0700 hrs. to care for the early rising combat and staff personnel, and complimentary to all this preparation is the fact that the generator providing current for the Headquarters' area will be kept running the entire night......

MISSION #79: AUG. 15, 1944

Today was D-Day for Southern France. I led the 1st box from our Group (487th), who were the first bombers over the invasion coast. Altogether there were 2,000 over the beachhead between 6:50 and 7:20 a.m., not counting the fighters and the troop carriers. There was plenty of action downstairs. Seven aircraft carriers, four battleships, etc. The Air Force had a corridor to fly and the Navy had a corridor for ships. The target for my box was gun positions at Cap Drammont, France. The Army was going ashore at 8:00 H. hour.

WAR DIARIES: AUG. 15, 1944

Squadron Leader Powell, RCAF, PRO was told to cover a combat assignment. This he fulfilled by flying with our formation this morning having as targets 8 different gun positions immediately preceding the first wave of troops landing in Southern France. Besides the all-consuming landing our air efforts supported, the day was also unique for these reasons: today the Group completed its 500th combat mission and today also marked 18 months of overseas duty for the Group. The Group also set a new high for itself sending up a total 1132 planes on 11 separate missions. The first ship took off at 0517 succeeded by 71 others. All these planes returned safely. At 1520 hrs sixty more planes took off with the Dreaded Avignon Bridge as target. The three bridges designated as targets were well hit but a price was paid. Three of the participating ships did not return and several of those returning were badly shot up..... Crews of the plane shot down over Legaro Bridge in Italy on June 3rd are reported to have safely returned to our lines. This fact was confirmed by the arrival of the gunner on this plane who was first interrogated by XIIth Air Force A-3, and sent here for three days to clear his affairs..... Some surprise was expressed by the news that Patton is in France in Command of an American Army.... Movie for the evening was a private affair for a selected few. The old projector was no longer serviceable, so regular movie was canceled. Late last evening a new one was picked up. Too late to show the regular movie and with the same curiosity as that to try on the first pair of pants, Corporal Becker, Group Special Service, called in a few of surrounding EM at HQ, told them to pull up seats and held a showing inside the Special Service tent. No one knows the picture's title, but they do know it involves saboteurs and seems interesting........

MISSION #80: AUG. 16, 1944

Flew as formation commander with 487th on road bridge at Livron, France. Saw lots of action when we crossed over the new beachhead.

WAR DIARIES: AUG. 16, 1944

The problem of keeping unauthorized transients from the S-2 Office has by no means been solved. A sign hanging above the doorway lapping low enough to scrape the head of those entering has still to attract the interest necessary to even stop to read what it imparts. Principal preoccupation is to eliminate the continuous disappearance of classified periodicals, to prevent as far as it is humanly possible the leakage of target information before it reaches the wrong persons, to avoid the unnecessary confusion caused by the office being used as a meeting point for all drifters. Major Johnson's statement that we have sacrificed mobility for extensive accommodations as a contributing factor is certainly with its share of truth for had the office been retained on its skeleton level as in Africa rather than a spacious real-estate office all this lingering would quickly disappear.... Major Kisselman jumped again on Captain Hangar's neck for necessary supplies and again Captain Hangar

suggested the writing of more requisitions for all their dubious worth.....
The new projector, put into use before the Group Movie for the first
time, was found deficient in one respect, its sound system is too weak
for the area it must cover. Steps are being taken to remedy the situation,
but so far as last night's movie few if any had the patience to see its end.
Nothing was comprehensible.... 340th hardball team walked off with
another victory last night beating the 41st Engineers by a 5-2 score. It is
surprising the attendance these games draw. The entire diamond is
surrounded with parked vehicles coming from any number of places to
see the game. A fair estimate places the average attendance at 1,000....
Cigarettes are now sold for $10.00 a carton in Sicily. A good many of the
boys are picking up their share of small change through this media. This
morning's missions, the Livron South Railroad bridge and La Pousin
Road bridge, both in close proximity to the other in Southern France
were successfully hit. One span of one of the bridges had completely
collapsed.... News of the new beachhead still is fragmentary with little
concrete facts given. More detail accounts should soon follow......

MISSION #81: AUG. 19, 1944

*Flew as formation commander with 487th on rail bridge at Orange, France. Lots of
heavy flak.*

WAR DIARIES: AUG. 19, 1944

The boys today made up for the two day let up by putting two compact
patterns across the center of the Orange RR bridge and dropping the
center span of a suspension road bridge at Montfaucon. A XIIth Air
Force PR release for today gives some indication of the efforts
undertaken by the B-25s in this recent landing in southern France. It
reads as follows, "For eleven days prior to the invasion of southern
France, B-25 Mitchells of Tactical Air Force attacked targets in the
landing areas to drop more than 25,000 tons of bombs in the softening
up program. During this period, the Mitchells put 1,129 planes over the
Axis-held territory in 68 combat missions, before August 2, the B-25s
have never bombed France but had concentrated on Italian targets for
more than eight months. Since their first attack and through D plus one,
the Mitchells have flown 100 missions, flown 1,603 sorties and dropped
3, 233 tons of bombs."   310th Band (accomplished but tiny) 321st
personnel with several colored boys from the infantry outfit near the two
organizations pooled talents and put on a class "A" variety skit show.
Many of the men had considerable vaudeville, experience or so it

seemed, for the performance was well above the ham and egg type of stage shows normally presented. For one thing, being themselves GIs, they knew precisely what commonplace subject matter to choose for their skits. As for example the jostling one receives when riding the back end of a GI truck or humorous sidelights and tribulations of an aerial engineer, various pantomimes were also very effective. Two things made it particularly enjoyable, the lack of flagrant vulgarity and the total absence of the 100% Zone of the Interior patriotism.... The HQ softball team won a heart breaker from the 1068 ordinance team by winning 1-0. Both teams made but one hit a piece with the winning tally coming in only as the result of four successive walks.....

MISSION #82: AUG. 21, 1944

Flew as formation commander with 487th on Rail Bridge at Parma, Italy.2

WAR DIARIES: AUG. 21, 1944

The still crisp morning air was shattered with the shrill blast of a policeman's whistle. The purpose of the whistle was not intended to direct traffic, however, as it was to disentangle each and every soul in the HQ's area from the firm grasp of Morpheus. And so it shall be now and ever after, so says Major Bennett, Group Executive Officer, that the C.Q. at HQ, will visit each and every tent ever morning at 0700 hours using the irritating call of a whistle to bring the men back to this world. The objective apparently seems to have the men at breakfast sometime before noon and have them report to work sometime before late afternoon..... The HQ's EM club declared a dividend of five dollars for all members who had invested the initial five dollar membership fee. Since the club has continued to make large profit rather than have any one or two do all the work as had previously been the case, four men, volunteers, will be given a total of 80 dollars a month to tend bar and keep the place neat... The blood curdling rivalry existing between the various sections of Headquarters has now found expression in softball games between sections. S-2 vs S-3. Corner lot ball games seldom produce any more vociferous arguments or more shouting. The main surprise of the evening was the presence of old frozen faced, 60-year old Major "Daddy" Summers. The Connie Mack of the 340th came out on the ball diamond ready for action with pant legs tucked in, sweat shirt on. It was well beyond the realm of one's imagination to watch this otherwise unnoticed soul attract so much attention by his astute deftness in batting and fielding. The men were convinced that Daddy Summers had played ball before...... Something out of the pages of Fenimore Cooper's "Deerslayer" came into the office last evening which had the men without words for the moment. Looking very formidable, helmet, green paratrooper's suit, German pistol on his belt, tommy gun slung over his shoulder, Glider pilot, 2nd Lt. Smith had stopped in to obtain a lift to Italy. Apparently, he was the real EM Club on Corsica.

McCoy for he had seen action in northern and was now returning from southern France where he had also made a landing. Our day's operation took us back to northern Italy with the Parma W. RR bridge as target. Two separate missions of 18 ships were sent after the objective. The bridge had been under extensive repairs since our effective bombing of it early July. Jerry was probably back at work this morning beginning his repairs from scratch for the bombing really made mincemeat of the bridge. Pvt. Henry F. Jolley, court-martialed twice in the last year, now under Major Johnson's guiding hand took the week's prize by returning early from his pass to Catania.

MISSION #83: AUG. 23, 1944

Flew as formation commander with 487th on railroad bridge at Avignon, France. This is considered the roughest target for anyone in the theater. We had an element of chaff and frags and 4 elements of frags that we dropped on the gun positions. It must have had the Jerries in their slit trenches, because we didn't get a shot.

WAR DIARIES: AUG. 23, 1944

On morning mission 8/10 cloud coverage with rain overhead prevented proper observation of target consequently all bombs were returned. Afternoon was a mixture of affairs with one mission sent to effectuate a road block North of Turin where two German Divisions were seeking a retreat through the little town of Settimo. The orders were to knock out the "Ponte San Marion [Martino]" crossing the Po Valley at this point. Both the bridge and the town were well covered. The second afternoon mission was Avignon Bridge. Already five ships had been lost on two missions to this attempt to avoid as much loss as possible a staggered flight was sent with three ships carrying frags and chaff with every two boxes. Flak was reported as Heavy, Scant, Inaccurate seeming to indicate the effectiveness of our plans, or a withdrawal of German guns. The bridge was not hit. Corporal Capawanna, now Sergeant, shot down over Avignon on our first attempt is reported safe in Naples and sent to XXIIth Air Force A-2 [sic. S-2?] for interrogation and disposition....

Flying Photographer and staff Sergeant in this Group, overseas 16 months completed fifty missions was sent home on furlough. Before leaving he gave instructions to the postal clerk to hold onto his arriving mail till his return. On arrival in the States he found his wife madly in love with some Lt. Commander. Only 300 of the 5000 dollars sent home in this period accounted for and his wife asking for a divorce. Married shortly before leaving for overseas he had but the oft repeated words of love and endearment upon which to base his belief in her fidelity. The divorce was granted. Upon opening and reading the accumulations of letters his wife had sent in the time required for him to reach home she still professed her great devotion to him and her great joy in anticipating his return home after so long an absence...... Paris liberated, and Romania becomes our new great forward thinking ally. The news was not taken with equanimity for it proved the needed justification for the usual drunks to add a couple more under their belts. Pvt. Harry Ford Jolley, through the intercession of the F.B.I. finally received word his wife is still alive and was notified of her present address. 13 months of patience finally awarded..... Movie "Smith of Minnesota." The writer being a former Michigan man felt several points difficult to follow or swallow......

MISSION #84: SEPT. 13, 1944

Flew with the 488th as formation commander on railroad bridge at Peschiera Del Gorda, Italy.

WAR DIARIES: SEPT. 13, 1944

Brigadier General Cragie, Commanding General of the 63rd Wing, always interested in the work of medium bombers, especially in the light of the fact that his fighters are constantly being called upon to act as our escort, flew with the Group today on its missions over two RR bridges in the Po Valley. The bridges were the Solignana and Piacenza RR bridges. Though both bridges were well covered with concentrated patterns, the General only able to be at one place at one time, witnessed the destruction of the Piacenza bridge.... After the mission Colonel Chapman had General Craigie and General Knapp as guests at dinner. Apparently enjoying both meal and mission...... As added attraction at the EM Club last evening they had hard eggs, but unfortunately no beer..... Lt. Wilbur Snaper, recently commissioned 2nd Lt. from S/Sgt. and who is the apparent heir to the throne at Group S-2 pending the departure of either Captain King or Major Kisselman and who has shown a very keen interest in this work, received a call from Major Field informing him that he would be placed on orders as Ass't T.J.A.[The Judge Advocate] for the Group. Lt. Snaper, though not a lawyer has had some experience in court room work as a civilian attempted to explain to the Major that he did not feel qualified for the job. Major Fields having no other choice nevertheless placed him as Ass't. T.J.A....... 306th alert to move again called off..... Group also flew several planes on a nickeling mission over Bologna and Ferli dropping the weekly edition of "Frontpost."

MISSION #85: SEPT. 18, 1944
Formation commander with 488th on troops at Rimini, Italy.

WAR DIARIES: SEPT. 18, 1944

The Italian K.P.s who have been with us since we were located at Pompeii A/D [Airodrome] have been ordered to pack up for they are being returned to their homes. A recent directive from XIIth Air Force made the Group's choice on sending them home a matter of following orders..... Mission again was gun positions and troop concentrations near Rimini. The first box failed to bomb because of clouds, the second formation with the same area as target successfully covered the area with frags and 100 lb demolitions bombs. A scheduled afternoon mission was canceled due to weather...... Tenacious, stubborn German opposition finds them still defending the Rimini sector. Again the men's spirits fall into one of suspended animation. Allied propaganda scheme, though so necessary to the morale of the men, has a substantial reverse effect discrediting entirely whatever belief the men place in its publicized news reports. From the fall of Cape Bon, this propaganda has constantly made it appear that German collapse was imminent, that their industry and fighting power was destroyed and with each successive German withdrawal this sort of tripe has been intensified. Now it is one and one half years later and those same Germans on the brink of collapse 18 months ago are still fighting harder than ever before...... The headquarters residential area takes on the appearance of a rushed defense housing of every tent busily converting them into petit joli cottages. The first signs of all came in a torrent last night. Many who had not taken flood-control measures are neck deep in mud with all their personal belongings well soaked in mud.

MISSION #86: SEPT. 24, 1944

Formation commander with 488th on target at Piazola railroad bridge. Went all the way to the target area but there was a complete overcast.

WAR DIARIES: SEPT. 24, 1944

Targets were the Piazzola RR bridge and the Cittadella diversion constructed over the river bed to circumvent the bridge destroyed in this area... Formation failed to drop because of weather. Movie for the night was "Shine on Harvest Moon" with the buxom Ann Sheridan and Dennis Morgan with story being based on the romantic theatrical success of Norwood and Bayes team of the late nineties and early 1900s. These seem to be well enjoyed.

MISSION #87: OCT. 1, 1944

Was command pilot with 488th on Piacenza railroad bridge. Lots of ack-ack and had our left aileron shot out.

WAR DIARIES: OCT. 1, 1944

Men commenced their period of winter hibernation today changing into the re-issued O.D.s. The day was clear and beautiful besides being warm. The O.D.s were a bit uncomfortable in the afternoon hours. M/Sgt. McElroy, Group Sergeant Major, was detailed by Colonel Chapman to be in charge of putting a tin roof on the EM Red Cross club and boarded up the windows for purposes of converting the club into a theatre in the evenings...... Three missions today took us to the Piacenza area with barracks, a gas plant and one of the RR bridges as the targets. All three were satisfactorily bombed. A fourth mission took us to the North Magenta RR bridge and here again the boys did right well for themselves.

MISSION #88: OCT. 12, 1944

Flew with 489th. Target: ammo storage at Bologna, Italy. It was a coordinated attack with all of the Mediterranean aircraft on the same area.

WAR DIARIES: OCT. 12, 1944

The "pancake" plan in direct support of the advancing 5th Army troops was executed this morning. As with the attempted Cassino Offensive of last February 14, this Group was the first over, followed by other Mitchell Groups, Marauder Groups (42nd Wing) Fortresses and Liberators. 5th Army observers forwarded a cablegram to the effect that it was a grand show. 1st Lt. Howdy, still with us and who flew in a camera ship stated he would have quite a good bit to tell his men when returning to the ground forces and especially his own organization who is in the vanguard of this present action. Our particular target in this spectacle was the Casalecchio Bivouac and stores area near Bologna........ Many bundles of Target Charts covering southern Germany, Austria, Hungary have arrived. Possibly never to be used, but which must nevertheless be kept on had to cope with whatever eventualities may arise.......

MISSION #89: NOV. 4, 1944

Flew with 486th Sqd. as formation commander on bridge at Villafranca, Italy. The 489th lost one ship and the 488th had one ditch, in which the pilot and tail gunner were lost.

WAR DIARIES: NOV. 4, 1944

Finally the monotony of doing nothing was broken as the group successfully attacked bridges at Casale Monferrato and Villafranca d'Asti. Two ships were lost when the lead navigator took the formation over Allessandria. One plane was shot down there [Lost were: Donald H. Rossler, James R. Gittings, Sidney L. Newman, Sgt. Mallicoat, Henry W. Harris, Chester E. Corle] and another was forced to ditch at sea. Of the crew in the latter plane the pilot was drowned, although his body was recovered, and the gunner is missing. [The lost men were 2nd Lt. Bryon D. King and Sgt. James A Burger] The other four crew members were rescued and brought back safely.

MISSION #90: NOV. 9, 1944

Flight leader for the Group with 488th Sqd. Had Major Cassada as my copilot, Chief Myers as my bombardier, and Maj. Nash as my tail gunner. Target: rail bridge at Tomba, Italy near Austrian border.

WAR DIARIES: NOV. 9, 1944

Captain Anthony J. King, assistant S-2 officer, took off on a five-day pass. Gossips are wondering whether he'll try to hop up to England, where he has an ailing grandparent. That country and France are, of course, in the European theater of operations, while we are in the Mediterranean. Standing orders prohibit our personnel from junketing up there, but the stunt's probably been done before.... There is still no word on the 486th plane that disappeared over Elba four days ago while en route to the target. Flying with a box from the 486th squadron, it simply wasn't there when the formation came out of a cloud bank... Two missions were scheduled today, but only one took off, and that was abortive because of clouds over the target. (Tomba bridge). Latest Malta quotation on English scotch whiskey is $11.00 per quart.

MISSION #91: NOV. 17, 1944

Flew as command pilot with the 486th. Target was road bridge at Faenza, Italy. It was a close support mission. Our left engine kept cutting out above 8,000 feet.

WAR DIARIES: NOV. 17, 1944

Thirty-six planes went out again today to tackle the two road bridges at Faenza on the 8th Army front, but again did little damage to the structures, although the approaches were hit.

MISSION #92: DEC. 2, 1944

Flew as 1st pilot on a mission to drop equipment behind enemy lines for the partisans. Had 4 P-47s for escort and dropped equipment at town of Bardi, Italy. Flew over the town on the deck and the people were there at the appointed time to pick up the stuff. 11 hours 40 minutes.

DEC. 7, 1944: PROMOTED TO MAJOR

WAR DIARIES: DEC. 7, 1944

Major Louis Keller, assistant group operations officer, has been transferred to 57th Wing, where it is rumored he will fly a P-51 on weather recce work for wing operations. His place as assistant operations officer will be taken by Major Fred Dyer. It is rumored around 57th Wing that the entire wing will lie around Corsica until the European war is over and possibly some months longer. There were no operations today owing to weather.

MISSION #93: DEC. 9, 1944

Flew as 1st pilot. Target was to drop equipment behind enemy lines about 15 miles north of La Spezia. Something must have gone wrong because we circled the area for one hour but did not get a signal to drop. Had 4 P-47s for escort.

MISSION #94: DEC. 13, 1944

Flew as 1st pilot. This was a supply mission behind the lines near Borgo Val Di Toro. Found the place OK and men ran out to mark the spot on the first pass-over. We dropped six bombs full of equipment and they hit right on the spot.

WAR DIARIES: DEC. 13, 1944

[No entry for this date.]

MISSION #95: DEC. 22, 1944

Formation commander with 489th on target at Lavis on the Brenner Pass line. Ran into lots of ack-ack and picked up quite a few holes. Really was a cold ride for four and a half hours.

[In between these missions George was working in Group Headquarters planning the missions, although not flying all the time.]

WAR DIARIES: DEC. 22, 1944

Our first mission in seven days brought the formation home all shot up and scattered all over the sky. The Lavis railroad bridge was the primary target but the formation couldn't find it early enough after hitting the initial point. One box bombed and cut an alternate bridge target, Chiavari. The accurate flak damaged two aircraft so badly they had to make emergency landings at fields in Italy, and three of our B-25s made emergency landings at Alesan.... Throughout the day the weather here was cold and blustery, and yesterday it rained killing operations. Apparently we're catching the worst of the Corsican winter weather right about now... A very large collection of candy, garnered from the PX rations of officers and enlisted men in voluntary contributions, is being built up in the public relations office. Two large G.I. equipment cases in the office are loaded down with 25 cubic feet of hard candy, peanut bars, chocolate bars, and packages of gum. The kiddies of the little town of Cervione will go wild when they get all that.

MISSION #96: DEC. 30, 1944

Flew as flight leader with 488th Sqd. on target at Calliano. Railroad bridge #1. Lots of ack-ack and when I broke off the target I looked back and had to laugh at Jinks, who was heading the 2nd box.3 Capt. Myers (Chief) was my bombardier and he hit the 120-foot bridge dead center from 12,400 feet.

WAR DIARIES: DEC. 30, 1944

The Calliano rail bridge #3 was hard hit today by the group, and an alternate target, Palazzolo rail bridge, probably escaped damage. A mission to Calliano #2 damaged only the tracks near the bridge, while Calliano #1 was probably hit. The alternate target for these missions, Crema railroad bridge was possibly damaged. Flak was troublesome on the missions, and Sgt. Raymond J. Alwood of the 486th, a gunner, was seriously hurt.

MISSION #97: JAN. 17, 1945

Flight leader with 488th Sqd. on target at Calliano railroad bridge #1 on Brenner Pass line. Chief Myers again hit the bridge, which had been repaired since we last hit it. Was officially assigned to Headquarters as of Jan. 10th.

WAR DIARIES: JAN. 17, 1945

M/Sgt. Tom Lennon, section chief in group operations, is in the hospital with a case of near-pneumonia, and his tent-mate S/Sgt. Ed Lorenz, also an operations clerk, is under observations for stomach ulcers.... Heavy rains yesterday knocked out operations, but today the 340th got bombers out to the Calliano bridges on the Brenner line and also Rovereto. Latter target was missed, while Calliano #2 was hard hit, and Calliano #1 cut on the approach.

MISSION #98: JAN. 28, 1945

Flight leader with 488th on railroad bridge at Rovereto, Italy. Lots of flak and this was the hardest mission I've had to date due to cold and no oxygen

WAR DIARIES: JAN. 28, 1945

Today the San Michele railroad diversion bridge got an effective bombastic treatment by the 340th, but at Rovereto railroad bridge the boys missed by a hair. Yesterday our mission to Rovereto was abortive because of bad weather. Another attack on the Bressana bridge met with much success, however, several hits being scored on the bridge, apparently... Some of the enlisted men who have been with the 340th since the earliest days and whose appreciation of officers, after two years overseas, is almost wholly based on their evaluation of them as human beings, are getting much amusement from the present crop of newly arrived combat crew officers. Said one enlisted man: "Why don't they tell those guys in the States what it's like over here before they ship 'em over?" One (of the new officers) wanted to know where the Coke (Coca-Cola) machine was. Another guy wanted to draw his radio set from supply. He said they told him at the P.O.E. (Point of Embarkation) that all tents came equipped with radios.

MISSION #99: FEB. 21, 1945

Command pilot with 488th on the railroad bridge at Bressanone on the very top of the Brenner Lines. Jinks was flying as flight leader in the same ship. We had bad weather and finally had to turn around after a hard time through the weather. Got shot at from Vicenza.

MISSION #100: FEB. 28TH, 1945

With 488th plane as lead pilot of a 42-ship formation. Target was Salorna embankment on the Brenner rail line. Major Dyer was my copilot flying his 100th mission. He was acting as Formation Commander.4 I had a gunner, Sgt. Helferich, who was also flying his 100th mission. Capt. Myers (Chief) was my bombardier. Maj. Nash was my tail gunner. Had lots of pictures taken when we landed.

WAR DIARIES: FEB. 28, 1945

Majors Wells and Dyer, assistant operations officers for group, flew their 100th combat mission today, with S/Sgt. Robert L. Helferich, 486th turret gunner, who also reached the 100-mark with the mission. They led the attack in "7F", a 487th squadron B-25, which, like the personnel, was flying its 100th mission in combat. Although the attack – on the Salorno rail fill – was pretty much a bust because of a radio release malfunction, much fuss was made over the affair by public relations. The 9th Combat Camera unit took motion pictures of the crew as they stepped out of "7F" and a Stars and Stripes correspondent was on the spot to get a story for that familiar house organ of the military... A second mission, to the San Michele rail bridge, was very successful. Yesterday planes from the group went back to Spillembergo ammo dump and caused extensive damage there, photos revealing that 21 of 30 revetted huts in the target area were hit by our explosives. At Pagnacco another dump felt the weight of two patterns, and bridges at Motta di Livenza, alternate targets, were cut on the track approaches. Majors Wells and Dyer, with their 100 missions have flown a greater number of bomber missions in their single tour of duty than any other bomber pilot in the Mediterranean theater, medium or heavy bombers considered. It is believed that if they top 101, they will have flown more in one tour of duty than any such pilot has flown in two in this theater.

MISSION #101: MARCH 13, 1945

Formation Commander with 488th on Aldeno R/R. Fill in the Brenner Line. Had lots of flak and had an oil line hit in the right engine and had to go on single engine over the target. We had to drop out of formation but we managed to get back OK. This was the 1st medium bomber that has ever returned from the Brenner Line on single engine. We were on single engine for 2 hours. We were able to hold it at around 6,500 ft. We were shot at again crossing the Po Valley by 40 and 20 mm. We then had a hard time getting over the mountains between Po Valley and the coast.

WAR DIARIES: MAR. 14, 1945

Back from the States and 30-day leave are two more of our majors, both from the 488th; Homer B. Howard, operations officer, and Charles J. Cover, executive officer. Owing to a mistake in orders, Major Cover had been assigned to a squadron at a B-29 base in Nebraska, and was brought back only after much wire-pulling... The 319th bomb group, formerly of our wing, is reassembling at Columbia, S.C., our old base, and will fly A-26s, the new Douglas attack bomber that can carry a medium bomb load and is faster than the B-25... Yesterday's missions consisted of a successful attack on the Perca road bridge in the Po Valley and a nickeling mission in the Bologna area. On the missions today possible hits were scored on the two bridges on the Casarsa rail diversion, and also the Cittadella diversion bridge.

MISSION #102: MARCH 19, 1945

Command Pilot with 489th. This was the Group's 800th mission. The target was a R/R Bridge at Muhldorf, Austria. This should be the last mission I'll fly before going home for a thirty-day leave.

TO : LIEUTENANT GENERAL EAKER.

This souvenir booklet is presented to you by the 340th Bombardment Group, telling the story of our part from the time we began to soften Southern France through D-Day for the invasion, i.e: 2 August 1944 to 15 August 1944, D-Day.

The 340th Bombardment Group feels that this operation produced the most difficulties and hardest operational problems yet presented to be solved.

On the communication targets from 2 August to 11 August, 1944, strikes were recorded on every mission. On the gun positions from 11 August to 14 August, strikes were recorded on 13 out of 16 targets. These targets were attacked with only six-ship missions and while it is not a perfect score, it represents the highest striking ability of any medium group in the operation. On D-Day a maximum effort of 72 sorties on the invasion beach and gun positions netted all strikes with the exception of four six-ship missions which were abortive due to 10/10 cloud conditions over the target. A new daily record of sorties flown was established by the 340th that day with a total of 132.

We are particularly proud of the operation on D-Day. When faced with night takeoff and join-up, eight separate missions in the air simultaneously, and a 10/10 cloud cover over the field at 700 feet, the 340th was the only medium group which placed all it's planes in their correct formation, on course, and over target on time, exactly as briefed. This represents the maximum in effort, co-ordination, and discipline that this group has ever been called on to produce.

W. F. CHAPMAN,
Colonel, Air Corps,
Commanding.

Col. Willis F. Chapman presented this souvenir booklet, made from the silver aluminum skin of the B-25, to Lieutenant General Ira Eaker at war's end.

340TH OVER FRANCE

In the air operations which first preceded and then directly supported the Allied expulsion of the enemy from the coasts and hinterlands of Southern France, the 340th Bombardment Group (M) played an essential role. Pre-invasion strategy demanded that the enemy forces defending the coast of France from west of Marseilles to east of Nice be broken into isolated groups by destroying their means of communications. This objective entailed destroying road and rail bridges across the southern reaches of the Rhone and Var Rivers and along the western extremities of the Drome, which flows into the Rhone. Flying its first mission against these French targets 2 August 1944, the 340th Group registered effective strikes on every road and rail bridge on which it was assigned. Of the 10 bridges assigned crossing the Rhone, Var and Drome Rivers, the target at Avignon proved to be one of the hottest and best defended targets ever encountered by this group.

From 11 August 1944 to 14 August 1944, four days before "D" day in southern France, the 340th Bombardment Group (M) was ordered to destroy enemy coastal batteries on the assault areas in order to guarantee maximum protection to the invading armies. Flying sixteen six-ship missions against these minature pin point targets, the 340th Bomb Group scored an unexcelled record of thirteen strikes on sixteen targets despite the fact that the objectives were so minute as to be invisible to the bombardiers and to necessitate bombing them solely by reference points.

These highly accurate attacks on the otherwise lethal shore batteries undoubtedly saved many Allied lives when the landing craft of the assault troops finally swung into position on "D" day and discharged landing forces.

The climax however, came on "D" Day itself, 15 August 1944, when the 340th Bomb Group was called on to execute an operation which called for the maximum in skill, preparation, initiative, and "know how" even under the best conditions, to support the Allied landings in Southern France.

The early morning commitment for "D" Day, the attack orders for which were received late in the afternoon of 14 August 1944, demanded a pre-dawn take off and join-up of six missions of six aircraft each and two missions of eighteen aircraft each, totaling seventy-

two crews, the absolute maximum available. Of these, two flights of
eighteen each, 36 aircraft were directed against the Anthoer Beach
to clear it of enemy troops, mines, barbed wire and other hazards to
facilitate a speedy and safe landing of the 34th American Division.
Each of the six ship flights was directed against coastal gun positions along the southern coast of France. This was the first time
that medium bombardment had been called upon for night formation
flying in this theater, nor was there any moon for assistance. Each
of the targets assigned required careful and intensive pre-briefing
of lead crews, followed by thorough briefings of the entire "task
force".

Because operational higher command had, in the interests of
security, withheld attack orders until a late hour, staff officers
of the 340th group had to labor under great pressure to work out the
detailed plans of the mission before the pre-briefing deadlines.
As the first takeoff was scheduled for 0517 hours, the first pre-
briefing had to be at 0345, from which time on there followed a su-
ccession of overlapping pre-briefings, briefings, takeoffs and join
ups, all under the close supervision of the commanding officer.
To promote the maximum success of the missions, details of all missions were worked out with admirable thoroughness: flight leaders,
navigators and lead bombardiers were painstakingly briefed and their
assignments coordinated.

The prescribed course to southern France for all formations was
within a fixed corridor through which many other aircraft were being
funneled to their targets. No time tolerance was permissable so there
was no allowable margin for error. To accomplish the feat of getting
eight separate formations of seventy-two aircraft, briefed, off the
ground, formed up at their respective altitudes intact over the field,
and on course so as to arrive at their objectives at the proper moments, was a formidable operation.

To complicate an already difficult task, this group was con-
fronted with a solid overcast condition over the field and surrounding areas from 700 to 1200 feet. Since the low overcast condition
nullified the use of navigation lights, all crews were minutely briefed for the take off and initial maneuvers of the join up solely on
instruments. Without a break, takeoff continued at intervals of thirty
(30) seconds until all airplanes were airborne. Dawn began to break
as the formations maneuvered into position to go on course. Complete
overcast was encountered enroute from Corsica to the coast of Southern
France and an overcast to broken ceiling of clouds over the entire
target area.

Of the 72 aircraft, 24 experienced complete overcast over their
targets, and were forced to return their bombs to the base.

245

The remaining 48 aircraft were met by broken cloud conditions and a
very difficult haze beneath, yet all managed to register very effective
strikes on their targets. Two of these formations were forced to
make second approaches into their targets because clouds obscured
their aiming point on the planned axis of attack. Only perserverance,
initiative, and determination, coupled with superb airmanship, accounted for the destruction of these targets.

This effort of the 340th Bombardment Group (M) coupled with its
missions on the afternoon of the same day when an additional 60 aircraft were dispatched on successful attacks on the heavily defended
bridges at Avignon constituted new daily maximums for the group - 132
sorties, a record of unsurpassed and unexcelled achievement for medium bombardment. The crews, of whom all flew on this eventful day -
many of them twice - with few exceptions had never flown on night
combat operations. None had ever participated in night operations
requiring formation flying. They met their test without hesitation,
imbued with determination and spirit born of the knowledge of their
own skill and courage, confidence in their leadership, and a superb
display of air discipline.

To summarize this groups contribution to the invasion of southern
France: all primary targets, communication, were attacked without a
miss. The attacks on secondary target coastal gun positions, were
examples of "putting all your eggs in one basket" technique, in that
only one box of six ships using only one bombardier was used on each
of 18 targets. A score of over 81 percent hits, presents a record
which is unsurpassed and unexcelled. The culmination of this effort
on D Day, the perfect execution of a most difficult assignment in
the face of all obstacles and difficulties, clearly reflects the
fruits of intensive training, skillful planning, and rigid air discipline.

'White Litenin' was the
ship that I flew on the
invasion of Southern
France leading the
340th Bomb Group.
    - Bill Chapman

APPENDIX B:
TRAINING DURING COMBAT

When Col. Willis F. Chapman took command of the 340th Bomb Group he drew upon training techniques — both conventional and inventive — to achieve the needed goal of increasing bombing accuracy and effectiveness.

The following on-site booklet, put together during this period, depicts the making of a film by the 340th's 9th Combat Camera Unit, in conjunction with its Public Relations Unit, as it documents some of these mandatory training skills, creative methods of operation, and enviable results. Its footage also highlights the custom-made Squirt Gun Turret and provides previously unpublished film of Joseph Heller at work.

Appendix B: Training During Combat   231

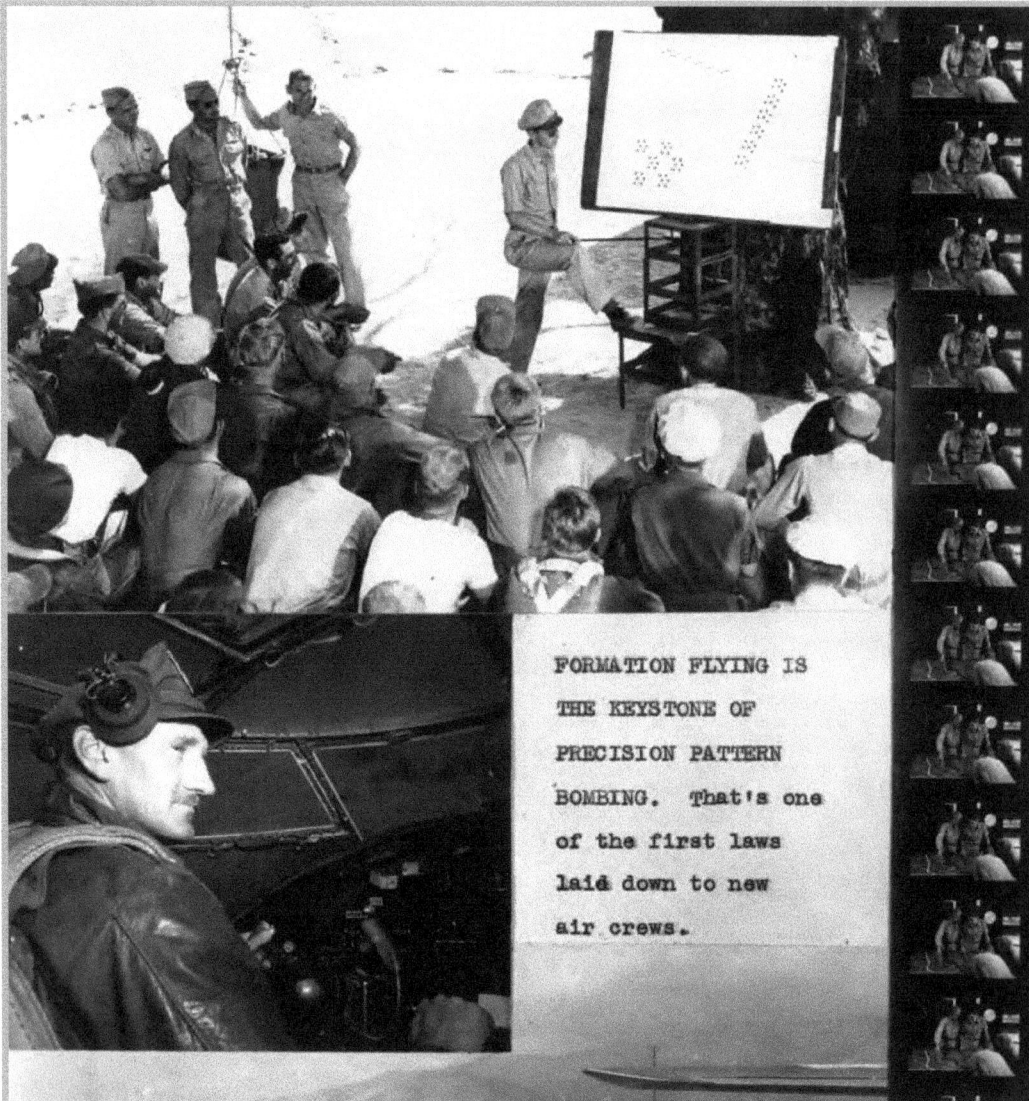

FORMATION FLYING IS
THE KEYSTONE OF
PRECISION PATTERN
BOMBING. That's one
of the first laws
laid down to new
air crews.

PROCEDURES and TECHNIQUES
which may or may not
have been practised
at the training
bases in the U.S.,
but which specifically
apply to their
operational area ,
have to be mastered
by the new crews.

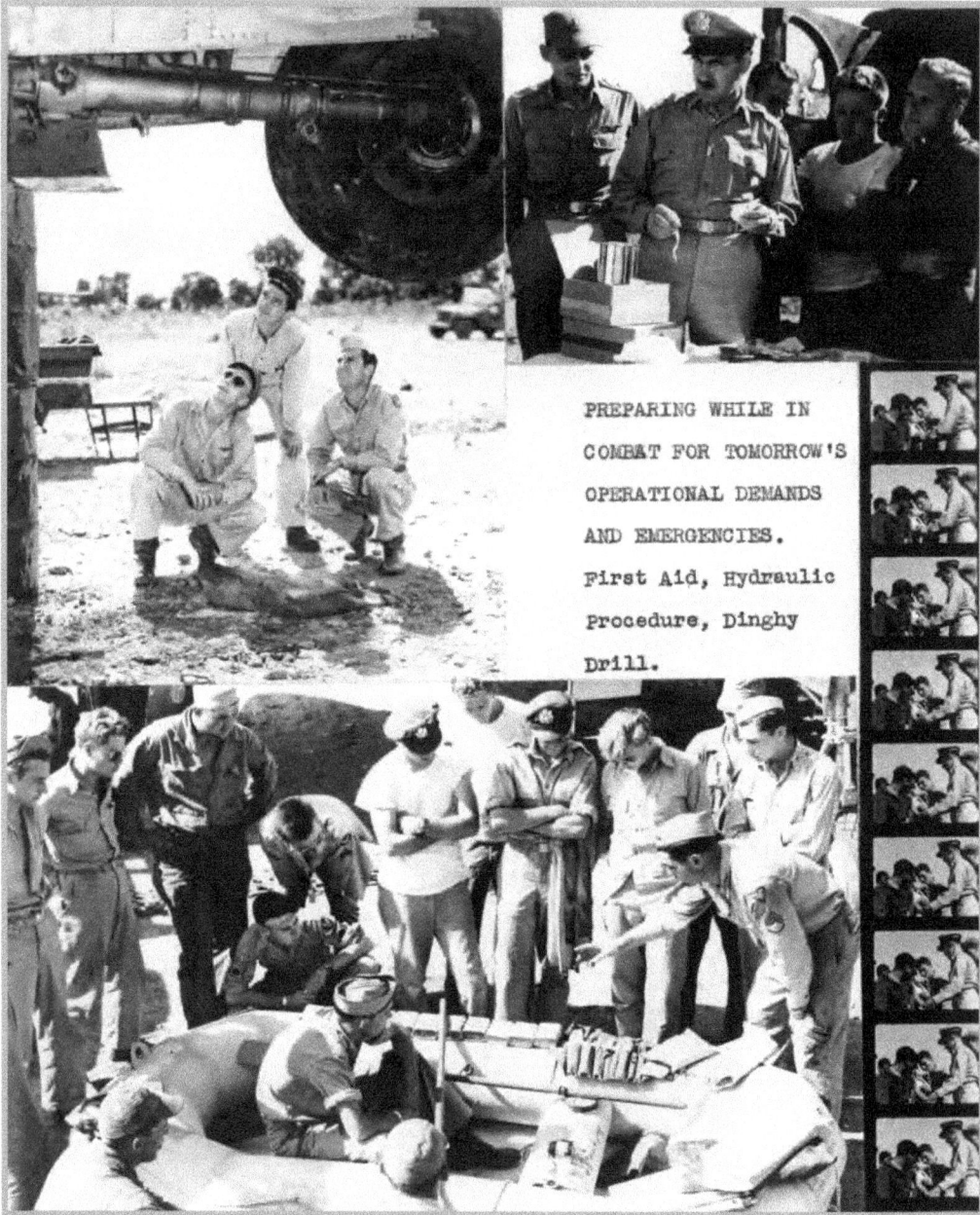

PREPARING WHILE IN COMBAT FOR TOMORROW'S OPERATIONAL DEMANDS AND EMERGENCIES. First Aid, Hydraulic Procedure, Dinghy Drill.

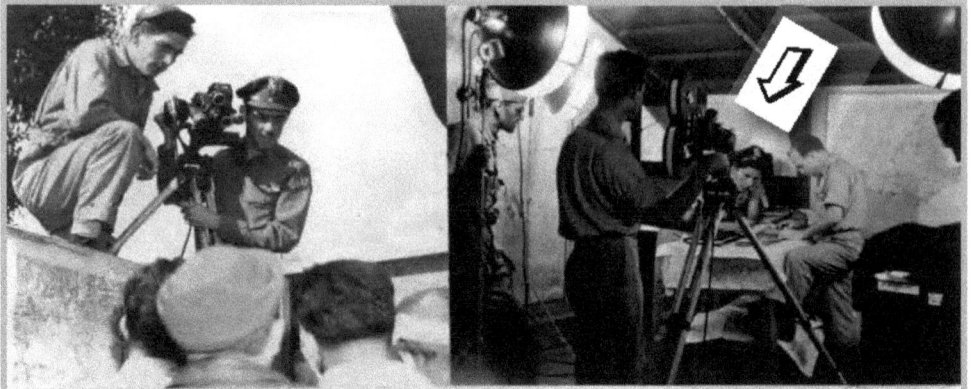

```
Photography..LT. W.T. BLUME
                                            9th Combat Camera
            SGT.V.R. KARNER
                                                Unit
Lights.......SGT.RAY  SEALOCK

            PVT.F.R. THOMAS

Script and
Narration....T/SGT. FRANK R. HICKEY
                                            Public Relations
                                            340th Bomb.Grp.
```

APPENDIX C
DISCOVERY! *A film – rare, priceless, and forgotten.*

The publication of *The True Story of Catch-22* started a chain of events that unearthed an historic WWII and literary treasure. Shortly after its publication a copy of this book was purchased by a person who, after reading the included manual, *Training During Combat*, was stunned to spy, in the credits, his father's name, Lt. W.T. Blume, 9th Combat Camera Unit.

2nd Lt. Wilbur T. Blume was a 340th photographer who had been chosen to spearhead the production of the film that manual had described. Son, Burton, immediately began a quest of his own to try to discover if the actual film had ever been made and, if so, was it still in existence.

Here the story takes on a life of its own. In 2014, deep in the recesses of the National Archives in Washington, DC, Lt. Blume's nine unedited reels were unearthed: 9 reels that had not been touched since the war's ending. While the finished edited product was not among them, these 9 reels contained copious footage from which the final film had cherry-picked.

The story unfolds.

In early 1944 the 340th Bomb Group was floundering under circumstances that had left it a bottom feeder in bombing accuracy. When Bill Chapman took command of this group, training methods, some standard and some unique to the 340th, were implemented without delay. Chapman's goal was always to encourage creative thinking among the men. He welcomed any new ideas. "Try it. It might work." These ideas created the Blue Bombs, the Squirt Gun Turret, and the chaff filled rocket. And work they did. As evidenced by the dramatic rise in the group's bombing accuracy, many of these unorthodox methods proved so successful that Group wanted them documented.

Lt. Blume, aided by Sgt. Frank Hickey, was tasked to put together a short documentary film of training undergone by these aircrews while flying combat. A script was written (Public Relations, 340th BG) around a combat replacement crew's arrival at Alesani Field on Corsica. It then follows these men on their induction into war's requirements and documents these unique training methods.

By the most unlikely, no, astonishing, coincidence, who was to appear in this documentary but Joseph Heller himself: he was cast as Pete who, like Heller, was a bombardier. The two protagonists were Bob, the pilot, and Pete, the bombardier. The script begins with Pete and his crewmates arriving on Corsica aboard a C-47 transport. From the moment these men exit that plane Heller is always recognizable for he had a habit of pushing his cap off of his forehead which exposed his identifiable wavy, black hair.

The script follows these men disembarking the plane, being loaded on to the back of an open truck, and then being transported to the headquarters area. As they mill about outside, here comes Group Commanding Officer Colonel Willis F. Chapman, along with their new squadron commander Major Randall Cassada (*Catch-22*'s Major Major Major Major), to welcome the new arrivals.

A crack in a parallel universe now begins to open as footage shows Colonel Chapman stepping forward and extending a welcoming hand to Lt. Joseph Heller, both sporting grins. Colonel Chapman greets Lt. Heller. Colonel Cathcart greets Yossarian. Here it began.

Heller's wide grin seemed to prophesy what was eventually to unfold when, years after the war ceased, true life allowed inspired fiction to blossom forth in the pages of *Catch-22*.[1]

Commanding Officer Colonel Willis F. Chapman
welcomes
Lieutenant Joseph Heller
to Corsica.

A full cutting script and narration that was discovered with these nine reels is here included exactly as it existed the day it was archived - untouched - unedited - directly from Blume's camera in 1944 as WWII raged.

SUB 2 583

NOV 7 1944

"T R A I N I N G   D U R I N G   C O M B A T

Cutting Script and Narration

PASSED FOR PUBLICATION
FIELD PRESS CENSOR

Narration by   Lieutenant Wilbur T. Blume
Ninth Combat Camera Unit

Technical Sergeant Frank R. Hickey
340th Bombardment Group (M) AAF.

T R A I N I N G   D U R I N G

C O M B A T   .

CUTTING SCRIPT AND SUGGESTED NARRATION

(Suggested opening is title "Training During Combat" superimposed on several shots of B-25 formations, building up to shot
of one large formation passing directly overhead, breaking up to
echelon peel-offs and landing. Material for this may be found in
rolls (Mags.) A-5, A-7, and A-8, Mag 12, Scene 97, and possibly
also library shots. This to be followed by Mag. 15, Scene 111,
to set the scene at a forward operational base.)

SCENES	NARRATION
	"Core and center of Army Air Forces tactical
	doctrine is the principle that training is a
	continuous process, even while our airmen are
	flying and fighting in the forward combat zones.
	Here in the skies above Alesan airfield, Corsica,
	crews of the veteran 340th Bombardment Group are
	returning from another successful mission. And
Mag 15 Sc 111	the men who fly these B-25 Mitchell bombers are
	veterans too, battle-hardened veterans wise in
	ways of aerial combat. Well prepared at training
	bases in the United States, they have won a
	brilliant reputation for precision bombing, and h
	added more luster to it through a thorough and we
	organized training and indoctrination program.

	Let's see what happens to a replacement crew when
Fade in	they arrive fresh from the States. Brought up to
Mag 13 Sc 100	the scene of action by the Air Transport Command,
101	they size up their new surroundings with interest
102	They're eager to go, this is what they've been
104 &105	training for, at last they're in combat.

Mag 3 Sc 118	Group Headquarters is their first stop. Outside
	the Operations building the interrogation is just
Mag 15 Sc 112	breaking up, and the hungry crewmen waste no time
	getting back to their squadron messhalls. (Mess-
	tents).

| Mag 15   Sc 113 | Well, well! They're traveling in fast company no |
| | if that sign means anything. |

Mag 15 Sc 113	Oh, oh, that's an eagle on that man's shoulder,
114	and he's probably the first Group C. O. Well, go
	ahead Bob, you're the first pilot

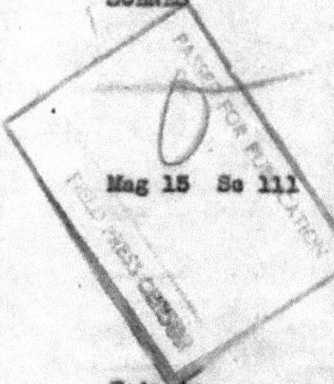

HEADQUARTERS
NINTH COMBAT CAMERA UNIT
APO   . N. Y.

TRAINING DURING COMBAT

SCENES

Mag 15   Sc 115

Mag 15   Sc 116

PASSED FOR PUBLICATION

Mag 3   Sc 117

FIELD PRESS CENSOR

Fade out

Fade in
Mag 5   Sc 25

Dissolve from closeup
of diagram to formation
shot, from mags or
library.

Mag a-3
(first three scenes)
(close scene with
formation flying away)

Mag 8   Sc 61

Mag8   Sc 64

NARRATION

"Right the first time! It's Col. Willis
F. Chapman(of Jackson, Michigan), and this
is Major Randall Cassada, commander of the
squadron to which our crew will be assigned.
He's their new boss now, and the questions
come thick and fast. They want to know all
the answers at once. How soon can we get
started? That's the big question - how soon?
Take it easy boys, there's another man over
here who can tell you more about that. This
is Captain Wells, fellows, Director of
Training. Pardon me, did you say training?
That's right and he's got lots of plans for
you. For the gunners workouts on various
training devices, aircraft identification,
and actual firing. For the bombardier and
pilots practice missions, navigation orien-
tation, formation flying. - and lots more.
Yes, chums, he has great plans for you.

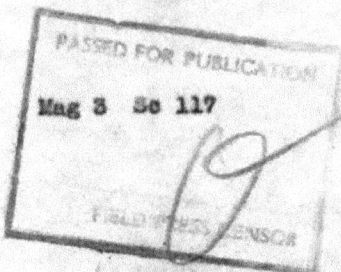

Formation flying is the keystone of precision pattern bombing. That's one of the
first laws laid down to air crews. Keep
those planes stacked in there close, and
you'll blanket that target. Watch your
element leader and your wingman. Don't
get lost on the turns, and remember, when
you get on that bombing run, hold her in
close.

Putting theory into practice these veteran
pilots and the new replacements work on
formation technique constantly. Bob's
flying co-pilot today, picking up some
pointers. Not too much different from what
they did in the States, but practice is what
makes a good bomb group better. These men
will be in training until after they've
flown their last combat mission.

Meanwhile our new bombardier, Pete, is
getting acquainted with his squadron navigation officer. A good idea because Pete
is going to have to do most of his own
navigation. Right now they're studying
maps and photo strips that reveal terrain
features. of areas their group will bomb in.

HEADQUARTERS
NINTH COMBAT CAMERA UNIT
APC   N. Y.

TRAINING DURING COMBAT

SCENES

NARRATION

Fade in
Mag 6 Sc 43

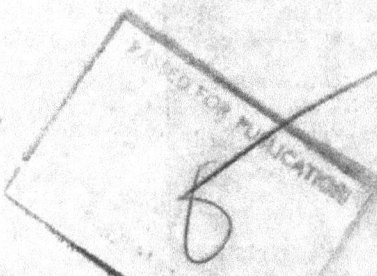

You are looking at one of the hottest aerial
gunners, undoubtedly, in the Army Air Forces,
even though he's never seen an enemy plane.
Trigger Thompson savvies all kinds of machine
guns, but his operations officer sent him sent
for a little workout on the squadron water gun
anyway. Here's how it works. You've got water
in the tank. It goes through that that tubing
and squirts out the jet below the guns. All
you do is aim the guns at the little model plane.
You press the triggers and give the plane a good
soaking - - we hope! Okay, Trigger, let's see
you try it. Say, this looks like more work for
the teacher than for the student./ You've got to
have   But,
          pressure in the tank to make the water squi
All right, Tally-ho! Drown the blighter. Let hi
have it!

Suggest cut-ins from
Trigger's face to model
plane revolving
(Mag 7 Sc 49)

What do you know! We missed it that time. Well
probably just cigar smoke in his eyes. A three
cent cigar at that, too. Try it again! (cut
from close-up of Trigger still shooting to other
training shots) .

Mag 3 Sc 119

Let 's see what's going on next door, while
Trigger's getting this thing under control.
No, this device is not a new mouse trap. It's
a gadget somebody dreamed up to make life more
interesting for aerial gunners. Those steel
rings represent the standard cones of fire used
in position firing. If a guy can just memorize
the relative positions of these cones, knocking
down a Jerry ain't so tough. And just so they
won't get too confused, some of the veteran gun
tell them what it 's all about. Say, I wonder
Trigger is getting along. From the looks of th

Mag 7 Sc 51
(long shot trainer
in operation)

he can use lots of practice. This thing isn't
easy as it looks. And oh yes, that's going to
more water pressure. Man the pumps! Jackson! —
These rookies!

HEADQUARTERS
NINTH COMBAT CAMERA UNIT
APO   , N. Y.

TRAINING DURING COMBAT

SCENES

NARRATION

Mag 3 Sc 12
Mag 10 Sc 77

Practice bombing missions, better than anything else, demonstrates to the new crews how a
group - their particular group - does the job.
Co-operation is stressed, not only between pilot
and bombardier, but also among every crew in the
formation. Here too, the old crews get a chance
to iron out past mistakes. Briefed as for a combat
mission, the individual crews go over their assignments before they take off. Not demos, but sand-
filled practice bombs are used. The small powder
charge in this bomb marks the point of impact and
gives the bombardiers a check on their accuracy.

Wipe from end
of above scene
to A-1, A-2, &
Mag 4, Sc 22
(aerial shots
practice mission)

Standard combat mission procedure is followed
including the use of evasive action up to the
initial point, where the bomb run begins. No
flak or fighters are expected on this mission, but
the tiny island offshore that serves as the target
is attacked with all the determination given to
an enemy key position. Bombs away, and every eye
strains for the tiny explosions that will indicate
success of the bombing. Now the formation wheels
for the home base, where, in a few hours, bomb
fall photographs taken by cameras in the planes
will tell the complete story.

Wipe to
Mag 8 Sc 66

Evidently the first bombs Pete has dropped for
his squadron were right in. With constant practice
that boy is going to make a leader-bombardier in

Mag 8 Sc 67

the not-too-distant fture. (Fade out).

Fade in
Mag 2 Sc 8

Knowledge of first aid treatment in the air often
spells the difference between life and death. And
the Flight Surgeon always has an interested ##
audience at his regular lectures. Today he's
explaining the use of the first aid kit carried
in the B-25 bomber. Somebody always has to be
the victim at these affairs - look whom they've
got here. Trigger's learning about all sorts of
things. HOW TO PREPARE FOR THE SPECIFIC EMERGEN-
CIES THAT MAY OCCUR ON THIS AFTERNOON'S MISSION 2
THAT IS THE TRAINING PROBLEM FOR THIS MORNING.

HEADQUARTERS   Page 5
NINTH COMBAT CAMERA UNIT
APO   N. Y.   TRAINING DURING COMBAT

SCENES   NARRATION

Mag 11 Sc 86   (phone rings, voice over phone, "This is
                                      ops calling. Briefing for plan "A" at
                                      1330//// hours. Got it?" Man puts down
Cut to   phone, picks up mike.
Mag 1   Sc 2   (voice over loudspeaker) "Attention all
                                      crews. Attention all crews.
Mag 1   Sc 1   "Men scheduled for Plan 'A' mission report
                                      to operations immediately. Briefing at 13
                                      (Announcement repeated immediately as men
                                      leave scene)

Mag 2   Sc 6   Well, our new crews isn't so new now. Look
                                      at those bombs on Trigger's jacket. They'v
                                      flown a few missions now, and had a chance
                                      see where training pays off.

Mag 10 Sc 82   (sound of engines starting)
                                      Bob is still flying co-pilot, even though h
                                      was a first pilot and flight leader ## in t
                                      States. His turn at the left seat will com
                                      however, after he gets a few more missions
                                      under his belt, and more training.
(Bob's plane taxies out, following by other scenes of B-25's taxyi
   and taking off. Material for these scenes may be found in Mag 9,
   Sc 76, Mag 11, Sc 91 and 92, Mag 12, Sc 93, Mag 14, Sc 106, Mag /
   and A-6, or library.) (Suggest use here of B-25's over target, in
   flak, bombs away, target strikes, planes hit by flak, planes whee
   off target - library

                                      It's a tense moment at the home base when
                                      the homeward-bound formation appears in the
                                      distance. Have they all come back? Are th
                                      badly shot up? As the echelons streak over
(Formations and   the field the men who stayed behind count t
echelons passing over;   returning bombers.
same references as
scene opening picture,   There's an echelon with only five planes in
also Mag 12, Sc 96 -   it - one missing
this last scene shows
a five-plane box)

Mag 15 Sc 109-10   No it isn't. There it comes, and he's just
                                      thrown a red flare. That means wounded
Mag 14 Sc 107   aboard.
Mag 12 Sc 95
Mag 12 Sc 94
Mag 1   At the far end of the runway ambulances wai
Combat Casualty   to pick up the casualty.
Sequence UNRELATED

HEADQUARTERS
NINTH COMBAT CAMERA UNIT
APO   N. Y.

Page 6

TRAINING DURING COMBAT

SCENES

NARRATION

Mag 1
Combat Casualty
Sequence (cont'd)

The Flight Surgeons waiting there will give the
wounded man emergency treatment before moving
him to the nearest station hospital. This man
has been severely wounded by flak in the shoulder
and thigh, and has lost much blood.   Good first
aid treatment in the air, however, has increased
his chances for survival. With the bleeding
already checked, the surgeons are free to admin-
ister plasma at once. The airmen witnessing
this scene know the value of their first aid
training.

(Fade out as
ambulance drives
away)

(Fade in
Mag 6 Sc 38

Here's a familiar scene, but what's that Link
trainer doing at a combat base? Well, they have
bad weather in combat too. Instrument flying
is a technique that has to be practiced frequently to be effective. Wonder who's inside
that thing? Well, our old friend Bob. A
little rusty from the looks of that chart, eh,
Corporal? But a little practice will remedy

Dark in there,
            isn't it?
(Fade in
Mag 14 Sc 108 #
(closeup of Bob
under hood in air)

that. Back to the instruments, Bob. A little
dark isn't it? (Fade out) Yes, it is dark, but
Bob knows where he's going with the aid of those
instruments. And just so he won't run into any
hot-dog stands, his co-pilot is taking every-

(out in shot of
single B-25 flying,
making shallow banks,
etc.

thing in. (interior shot showing both co-pilot,
and Bob under hood) You can come out from under
that hood hood now. Not bad, Bob, still
       right side up, and still over the home field.

(shot of B-25
banking and going
off in distance)

Say, it must be chow time!

Mag 7 Sc 53

That looks like a bomb trainer. I thought they
left all those in the States.& There's Pete
chasing around on a high chair trying to drop a
plumb bob on a penny. He and Bob have about ten
missions now, but they're still putting in time
on the bomb trainer. This type of training is
particularly valuable because it builds up the
pilot-bombardier work habit and turns two individuals into a team. The trainer itself makes
simulated bombing runs on the small target below
and the accuracy of the bombing is represented
on the chart that attendant is scoring. Did

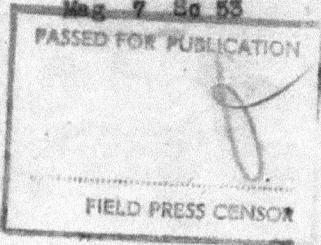

(Fade out as Bob
& Pete examine target)

they hit it, or did they hit it? Can you guess?

HEADQUARTERS   Page 7
NINTH COMBAT CAMERA UNIT
A   N. Y.   TRAINING DURING COMBAT

SCENES   NARRATION

Mag 16 Sc 122   Here is a variation of the water gun device
                                designed expecially for the upper turret gunn
                                Featuring two model planes, this device
                                enables gunners to practice meeting attacks
                                from various directions.
Mag 9   Sc 73   Elsewhere on the base gunners are at work on
Mag 9   Sc 74   other trainers. One of these is the waist gu
                                window mockup. It's valuable to the men who
                                man the flexible guns on the B-25's.
Mag 9   Sc 75   These veterans sure have it rough - always
                                pumping water for the rookies. My aching
                                back! Now they've got Trigger instructing.
                                Well, why not? After all the time he spent
                                on it, he ought to know something about it.

Mag 4   Sc 35   Certain targets are so situated that they are
                                difficult to identify on the approach. When
                                circumstances warrant a detailed model of the
                                target is prepared on the sand table, which
                                the lead bombardiers and navigators study car
                                fully before a briefing. Here the Group navi
                                gator is pointing out the most recognizable
                                features of the terrain.

Mag 5   Sc 32   At the regular briefing the rest of the crew
                                receive their instructions. The plan of
                                attack is discussed as well as the enemy
                                defenses likely to be encountered. When crew
                                leave this room, briefing and training have s
                                prepared them for that it is no accident when
FIELD PRESS CENSOR   they score a hundred per cent hits in the
Mag 11   Sc 87-8-90   target area. First Lieutenant Robert G.
                                Pagh (pronounced pag to rhyme with hag) Air-
                                plane Commander. That promotion, your own
                                airplane, and a first pilot's job speak for
Mag 11 Sc 90   themselves, Bob. Nice going.
Mag 11 Sc 91   And it looks a
(BY pulling out   if that gunnery practice paid off too.
on to taxi strip)

Mag 1   Sc 3   Methods of escaping the enemy if shot down
                                are the subject of frequent lectures. Note
                                Here the intelligence officer is passing out
                                packets containing escape aids. Replacement
                                crews are indoctrinated in escape and evasion
                                procedure before they fly their first mission

wheel lowering system.

Page 8

HEADQUARTERS
NINTH COMBAT CAMERA UNIT
APO   N. Y.

TRAINING DURING COMBAT

SCENES

NARRATION

Mag 16   Sc 120

The ditching of aircraft at sea and use of the life raft are periodically reviewed by all air crews. Each man is told exactly what he must do in this emergency. The importance of releasing the life raft immediately is stressed. The

Mag 16   Sc 121

instructor also points out the various items of equipment and provisions carried in the raft - and how to use them - - even the oars. Bend that back, Jackson, you're a long way from shore.

Mag 4   Sc 17

But the situation isn't so funny when a crew find themselves in the water. This one seems to be having difficulty turning it right-side up. Of course, in this case the boys are getting a little help from shore. And what a patient, understanding sort of fellow that instructor is. That's right, use that rope on the side! You guys in the back there, get under it and Heave! Hey, take it easy on those pants, that's his last clean pair! Okay, now you've got it; that's more like it.

Mag 3   Sc 10
Mag 3   Sc 11

Hydraulic failure can bring fatal results if the crew is not up on this emergency procedure. The ship being used here as a training mock-up came in on its belly several months ago when its hydraulic system was shot out by flak. By hooking up an external hydraulic system, crews are able to study and operate the emergency wheel lowering systems. Since this plane has been used as a school ship, no plane in this group has belly landed because of hydraulic failure.

Mag X-4

In combat the ability to fly on one engine is vitally important. Single engine operation is another skill periodically practiced by these bomber pilots.

SUGGESTED ENDING: A montage starting with bomb-fall plots falling into the boxes labelled "Combat Missions" and "Practice Missions" to denote training and combat going on at the same time. (Mag 8, Sc 68) followed by graphs on training and mission efficiency (data supplied in supplement hereto), accuracy of combat bombing (Mag 8, Sc 69-70, and perhaps cut-ins of various phases of training, accuracy of bombing as shown in Mag 8, Sc 69-70, with numerous 'bombs away' shots and bomb explosions superimposed on turning pages of bomb-fall plot book.   END.

PASSED FOR PUBLICATION

reservoir &
power unit

FIELD PRESS CENSOR

Precious few photographs had existed of Joseph Heller during WWII. This rare footage rectifies that while illustrating, through an artistic media, his daily life during the war.

Joseph Heller

- about to set foot on Corsican soil.

- being transported from plane to headquarters.

- is welcomed to the 340th Bomb Group.

-with the other new arrivals, meeting the340th commander, Colonel Willis F. Chapman.

- practicing with the Norden bombsite.

- plotting navagation

- scratching plans in the dirt.

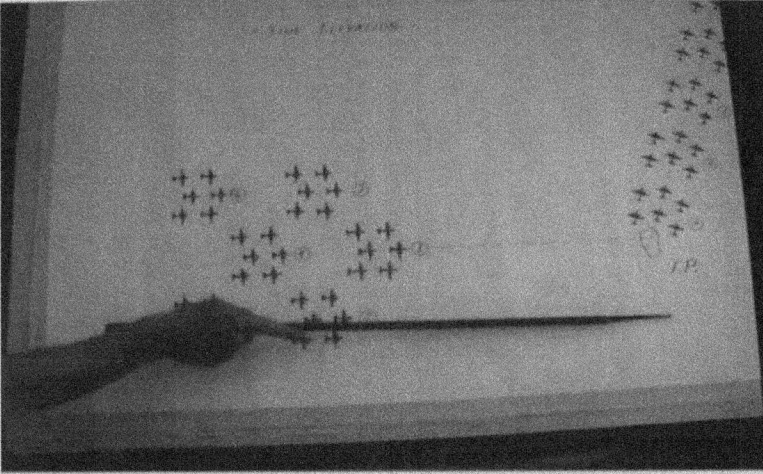

-attending the briefing. Boxes of 6 to the IP then turning toward the target.

PRACTICE MISSIONS

COMBAT MISSIONS

- with crewmates.

- with foot on top rail.

- *the bombardier*

-- in his plexiglas fish bowl.

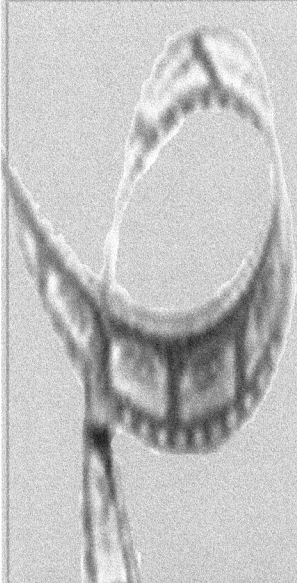

- observing the B-25's distinctive twin tail.

- flying formations so tight that friend's faces could be seen.

ACKNOWLEDGMENTS

To Casemate Publishing, for being *The True Story of Catch-22*'s first publishing house; Priscilla Reagan, my literary agent, whose whimsical nature has kept our relationship anything but ordinary; Linda D. Morehouse, Clarissa Moore, and Anne K. Kaler, whose excellent editing skills helped to burnish these pages; Bernard Argent, who skillfully mastered this DTP challenge; George L. Wells, dear family friend and invaluable consultant throughout this book; Willis F. Chapman, who valued and preserved his copious WW ll materials; Jason D. Reimuller, my most valued critic; and the men of the 57th Bomb Wing for their 'family' friendship and open hearted cooperation.

ENDNOTES

PART I
CHAPTER 1: GEORGE
1. IP=Initial Point: The designated spot where the aircraft would turn from its path and veer directly toward the mission's goal.
2. Some of the above Group history came from the personal files of Thomas A. Hetzel, a bombardier/waist gunner/photographer with the 340th who flew 70 missions.
3. Fred Dyer, personal correspondence with author, 1989.

CHAPTER 2: JOE
1. Playboy, June 1975, pp. 59-76.
2. Joseph Heller, "Catch-22 Revisited." Holiday Magazine, April 1967, Vol, 41, No. 4.
3. Playboy, June, 1975.
4. Susan Cheever, Special for USA Today. Interview with Joseph Heller.

CHAPTER 3: THE MISSION
1. Personal Stories on 488th Bomb Squadron web page, www.488thbombsquadron.com.
2. "Crippled Pilots Merge Assets To Bring B-25 In," Stars and Stripes, May/June 1944.
3. From author's taped conversation with her father.

CHAPTER 4: BILL
1. Colonel Willis F. Chapman, "How To Break In A New Group Commander," 340th Men of 57th, Vol. XXVl No. 2, Summer 1993.
2. Regarding chaff, submitted to the same newsletter, Men of the 57th, by an unidentified member, was this article about the effectiveness and great hazard to the crews of dropping phosphorous bombs.
3. Gill Robb Wilson, New York Herald Tribune, November 13, 1944.

CHAPTER 5: BOB
1. 57th Bomb Wing Newsletter, March 1980, p. 4.
2. Philip Lipshin, 445th.
3. Men of the 57th, March 1984.
4. Men of the 57th, March 1984.
5. Men of the 57th, Vol. XX. No. 4. June 1987, p. 1.

CHAPTER 6: GEORGE, AGAIN

1. George Wells, quoted on website of the 488th.
2. Bill Bancroft, "And Now—Reunion Talk 'Flyingest Outfit' Vets Relive WWII", Oakland Tribune, July 18, 1972.
3. Dialogue between Pat and George, 340th Bomb Wing Reunion, Virginia, October 2006.

PART II

1. Joseph Heller in interview by Susan Cheever, Special for USA Today, 1998.
2. George and Shirley Wells, personal conversation with author.
3. George Wells, personal correspondence with the author, November 7, 1988.
4. Joseph Heller, "Catch-22 Revisited," Holiday Magazine, April 1967, Vol. 41, No. 4, p. 59.
5. George Wells, personal correspondence with author, November 7, 1988.
6. George Wells, personal correspondence with author, November 7, 1988.
7. George Wells, personal correspondence with author, November 7, 1988.
8. George Wells, personal correspondence with author, November 7, 1988.
9. Forrest Wells, top turret gunner, telephone conversation with author, 2008.
10. Men of the 57th Bomb Wing, Fall 2002, Vol. XXXV, No. 3, p. 3
11. George Wells, personal correspondence with author, November 7, 1988.
12. Joe Heller, Interview by Susan Cheever, "From Coney Island to 'Catch-22,' a life story." Special for USA TODAY, 1998.
13. Excerpt from a Playboy Magazine interview of Joseph Heller, June, 1975
14. Heller, Playboy Magazine, June, 1975.
15. George Wells, personal correspondence with author, November 7, 1988.
16. George Wells, personal correspondence with author, November 7, 1988.
17. George Wells, personal correspondence with author, November 7, 1988.
18. General Robert Knapp, Men of 57th, Vol. XXII No. 4, December 1988.
19. Remembrances, The 57th Bomb Wing, Harold G. Lynch, 1980s
20. Heller, Playboy Magazine, June, 1975.
21. George Wells, personal correspondence with author, November 7, 1988.
22. Ben Kanowski, Seeks, *Catch-22* Characters Reunite in Albuquerque, December 13-20, 1975, pp. 16-17.
23. George Wells, correspondence with author, November 7, 1988.

PART III
CHAPTER 7: VESUVIUS SPEAKS

1. Men of the 57th, Fall 1999, Paul Gale.
2 . Axis Sally was a failed American actress turned radio announcer for Radio Berlin during WWll. This star-struck girl from Maine turned into the well-known and reviled disseminator of Nazi propaganda. With her American accent and knowledge of American music and culture, she would taunt US troops about infidelities at home and the horrible fates awaiting them on the battlefields. She tried to always cast doubts and heighten feelings of homesickness.

CHAPTER 8: I AM 7-K
1. Ward Laiten, "The Saga of 7-F," Men of the 57th, Vol. XXVIII, No. 1, Spring, 1994.

CHAPTER 9: CHAFF OVER THE TARGET
1. The Men of the 57th, Newsletter of the 57th Bomb Wing, September, 1977.

CHAPTER 10: WE LOSE THE QUESTION MARK
1. In this formation two planes and their crews went down. The Question Mark and Ship 8H. George, from his plane, helplessly witnessed 8H, the ship of his friend, "Red" Reichard, spiral down in flames. (See page 221, George's 37th mission.)

CHAPTER 11: SHORT SNORTER
1. Arthur William Tedder, 1st Baron Tedder of Glenguin, GCB (11 July 1890–3 June 1967) was a senior officer in the Royal Air Force and a significant British commander during the Second World War.
2. Spring 2000 issue of Men of the 57th. See Fall 1999.

CHAPTER 12: DOC FIGURES IN ONE OF MY FAVORITE WAR STORIES
1. Beryl Graubaugh, Men of the 57th, Winter, 1989.

CHAPTER 13: CHARLIE
1. Excerpt from article titled "Catch-22 Characters Reunited in Albuquerque," in Seeks, December 13-20, 1975.

CHAPTER 14: THE BRIDGE AT PARMA
1. Men of the 57th Vol. I, Issue XXXI, No. 4, Winter 1997, p. 16.

CHAPTER 15: PETRIFIED WITH FEAR
1. "A 50th Anniversary," 13 May 1944, Men of the 57th, Vol. XXVIII No.3 Fall 1994.
2. George Wells personal correspondence with author's father, Brig. Gen. Willis F. Chapman, 1999.

CHAPTER 16: THE SAGA OF 7-F
1. David Mershon, 487th. Men of the 57th, Vol. XXVIII No. I, Spring 1994.

CHAPTER 17: A LADY LEAVES THE 310TH
1. Frank B. Dean, 380th. Men of the 57th, Vol. XIII No. 3, January 1981.

CHAPTER 18: BRIEFING TIME
1. Olen Berry, 57th Bomb Wing Newsletter, Nov. 1979, p. 5.

INTERESTING FACTS
1. Dita Beard on Dita Beard (interview with Time correspondent Ted Hall, April 3, 1972)

APPENDIX A: GEORGE'S MISSION BOOK
1. Wells and Chapman had been given different headings.
2. This Bomb Group became known as the "Bridge Busters" because of their accuracy.
3. Jinks wrote George's parents and told them George had laughed at his getting shot at; George's parents told him he shouldn't do that!
4. They flipped a coin to see who would be the pilot, as they both wanted to be pilot. George won.

APPENDIX C: DISCOVERY!
1. At this book's publication the nine-reel footage, running almost 73 minutes, may be viewed in its entirety on:

   Youtube.com/TrainingDuringCombat

   From that footage, an 8-minute clip was made which includes that initial meeting of Colonel Chapman and Lt. Heller. This clip may be viewed online at the site:

   http://blogs.archives.gov/unwritten-record/2014/09/17/the-reel-Catch-22-pt-2-joseph-heller-and-training-during-combat/

GEORGE L. WELLS

HAS FLOWN HIS FINAL MISSION

JANUARY 18, 1919 – NOVEMBER 9, 2010

www.ingramcontent.com/pod-product-compliance
Lightning Source LLC
Chambersburg PA
CBHW040254100426
42811CB00011B/1257